Financial Return Risk and the Effect on Shareholder Wealth

Corporate Finance and Governance

Herausgegeben von Dirk Schiereck

Band 9

PETER LANG

Frankfurt am Main · Berlin · Bern · Bruxelles · New York · Oxford · Wien

Malte Helmut Raudszus

Financial Return Risk and the Effect on Shareholder Wealth

How M&A Announcements and Banking Crisis Events Affect Stock Mean Returns and Stock Return Risk

A Compendium of Five Empirical Studies across Selective Industries

PETER LANG
Internationaler Verlag der Wissenschaften

Bibliographic Information published by the Deutsche Nationalbibliothek
The Deutsche Nationalbibliothek lists this publication in the Deutsche Nationalbibliografie; detailed bibliographic data is available in the internet at http://dnb.d-nb.de.

Zugl.: Darmstadt, Techn. Univ., Diss., 2012

Cover Design:
Olaf Glöckler, Atelier Platen, Friedberg

D 17
ISSN 1869-537X
ISBN 978-3-631-62249-0

www.peterlang.de

Preface

Malte Raudszus has taken his first steps in corporate finance research back in the days before he graduated from European Business School (ebs) in 2007. Built on the learnings of a very well written seminar paper with a fellow student, his diploma thesis on "Risk behavior around convertible bond issuances" laid the basis for his ongoing focus on empirical studies investigating risk dynamics. Not surprising for a graduate from ebs he had various professional opportunities and proceeded with an initial career step in strategy consulting at The Boston Consulting Group (BCG).

In close communications throughout this time I was able to keep Malte convinced that a doctoral program in corporate finance would be a highly valuable experience and academic challenge to him. In the fall of 2009, I welcomed him back at the chair of corporate finance, now at Technische Universität Darmstadt, where I had routinely moved to in the meantime.

Until December 2009 Malte had yet achieved to finish a draft of his first study on corporate diversification with a focus on the construction industry. This enabled him to obtain admission to his primary academic conference, the annual meeting of the Academy of Economics and Finance (AEF), then held in Houston, Texas. Mid 2010 he submitted his manuscript to a top-tier international management journal where he is currently within the second round of the revision process.

Throughout the year 2010 Malte engaged in some cross-discipline research projects with a former fellow student from ebs. His spearing partner Jan-Willem Olliges was then enrolled as a doctoral student in the field of real estate economics with a former colleague of mine at ebs being Karl-Werner Schulte. As I had a first manuscript on my desk I identified significant publishing potential and recommended to get an American A-list researcher on board to challenge the study. Today I am delighted to say that this planting of seeds has carried nice fruit as a first publication of Malte, Jan-Willem and Glenn R. Mueller form Denver University in the Journal of Real Estate Portfolio Management has been achieved.

In terms of the academic learning curve Malte has climbed up high throughout his time of active research at ebs and TU Darmstadt. Yet in his diploma thesis he applied advanced volatility modeling techniques from the so called GARCH family. Probably not anticipated by him either he reignited on this topic when starting his doctoral studies and thus developed empirical volatility and risk analyses as one of his core fields of academic competence.

This dissertation as a compendium of closely related chapters but nevertheless individual studies underlines his way of performing continuously advancing research. As the first part sheds light on corporate diversification through M&A and how related risk dynamics affect shareholder wealth, he carries over the methodology to external events in the second part of the book. In studies three and four Malte individually scrutinizes the adverse implications of bank failures and bank bailouts in the recent financial crisis from 2007 to 2009. As he finds opposing return shocks he identifies the limits of the standard GARCH (1,1) model and moves on to the more advanced EGARCH which incorporates the so called leverage effect. As observed of the behavior of financial market stock return data, volatility reacts asymmetrically to positive and negative return shocks. As Malte applies the EGARCH in his final study he can simultaneously scrutinize the adverse bank events with an appropriate econometric foundation.

Throughout Malte's various conference attendances at the annual meetings of the AEF, the American Real Estate Society (ARES) and the Deutsche Gesellschaft für Finanzwirtschaft (DGF) he has received an amount of valuable professional feedback and insight into the research economy that is by far not common for a two-year doctoral student. Currently, three of his studies are under review in major international journals, and as it is the objective of any serious academic scholar, he is talking the effort to finally achieve the respective publications. In this light interpret the present book as a small plant that is yet growing and hopefully further refined and advanced versions of the present studies will find their way into a premier publication.

I am a bit sad to have lost Malte again to the world of business. As he had been focused on banking and M&A back in the days as a strategy consultant it is again not surprising that he recently continued to work in that field. Starting as of January 2012 he moved to Deutsche Bank where he currently supports and drives the ongoing integration of a major German retail bank. I still hope he returns to academia soon as he is not just a gifted scholar in financial econometrics but also enjoys the teaching. He has yet done very well stepping in for me when I was on travels and would serve as outstanding role model for our discipline.

In any case I wish him all the best for his future endeavors.

Prof. Dr. Dirk Schiereck

Acknowledgements

This dissertation origins not least from a seminar paper in finance which my good friend Alexander Ploch and I wrote during our final year of studies at the European Business School (ebs), Oestrich-Winkel. With his rigorous approach to econometric analysis he was a main driver of success to our joint work. Consecutively, we each wrote our diploma theses individually in the same field and I later continued on my academic path pursuing a PhD in finance. Therefore, I like to extend a special thanks to Alex as he clearly enabled me to take this way.

After graduation I took advantage of starting of my career at The Boston Consulting Group, even though I had yet committed myself to a doctorate program at ebs. I am very thankful to Dirk Schiereck, then Professor of Banking and Finance at ebs, that he showed himself obliging in extending another offer for a doctorate program to me, after he had moved to Technische Universität Darmstadt (TUD). Throughout my dissertation time he has been a great 'doctor father' and I am especially thankful for the countless times Professor Schiereck took the burden to provide me with valuable advice, comments, criticism, and suggestions.

Special thanks go to Ramit Mehta and Jan-Willem Olliges as they contributed significantly to three of the five studies this dissertation book comprises, to Professor Glenn Mueller form University of Denver, USA, and to various anonymous referees who put considerable effort into achieving publication in major academic journals. I am also grateful to my doctoral colleagues at TUD who engaged in insightful discussions and helped refine my thinking. In particular I want to thank Julian Trillig, Mehtap Ölger, and Mark Mietzner for their patience and effort in discussing econometric issues and technical transformations.

This book could hardly have been written without the support of my family and friends. Intensive studying, literature analysis, and econometric modeling need counterparts to keep up the mood. Therefore, I like to extend extra special thanks to my parents, Barbara and Frank, supporting me again for the time being. Another distinguished thanks goes to the 'student horse riders' at TUD who enabled me to pick up a unique sport and dive into a thrilling community that is very ambitious but also dedicated to a good bit of fun.

All remaining errors of this dissertation are mine alone.

Malte H. Raudszus

To my families.

Content Overview

Table of Contents

List of Tables

List of Figures

List of Abbreviations

ABS	Asset-backed securities
AER	Aero [industry]
a.o.	among others
ARCH	Autoregressive conditional heteroscedasticity
BEVER	Beverages
Big Four	Four largest breweries worldwide: Anheuser-Busch Inbev, Heineken, SABMiller, and Carlsberg
Bil.	Billion
CAPM	Capital asset pricing model
CGI	Consumer goods industry
CON	Construction
CSI	Consumer services industry
DEF	Defense
DG	Drug
DS	Datastream
DW	Durbin-Watson [statistic]
e.g.	exempli gratia (for example)
EGARCH	Exponential GARCH
ELE	Electric
ENG	Engineering
EQP	Equipment
Fed	Federal Reserve Bank
FD	Food
FDIC	Federal Deposit Insurance Corporation
FIHI	Financial Institution Health Index
FSI	Financial services industry
FTSE	Financial Times Stock Exchange [Index]
GARCH	Generalized autoregressive conditional heteroscedasticity

GEN	General
GDP	Gross Domestic Product
GDS	Goods
HOU	Household
i.e.	id est (that means)
IND	Industrial
KS	Kolmogorov-Smirnov [test]
LEI	Leisure
LOLR	Lender of last resort
M&A	Mergers and acquisitions
MAT	Material
M-GARCH	Multivariate GARCH
Mil.	Million
ML	Maximum likelihood [estimation]
NA	North-American
NAREIT	North-American REIT
OLS	Ordinary least squares [regression]
PAR	Parts
PROD	Production
REIT	Real Estate Investment Trust
RET	Retailer
S&P	Standard & Poor's
SER	Services
SIC	Standard industry classification
SDC	Security Data Company
SUP	Support
TARP	Troubled Asset Relieve Program
TRA	Transportation
TUD	Technische Universität Darmstadt

Txn	Transaction (M&A)
U.S.	United States [of America]
USD	United States [of America] Dollar
Wx	Wilcoxon [test]

List of Symbols

Greek symbols

α	Multivariate regression intercept
β	Systematic risk factor (beta)
γ	Coefficients to GARCH and EGARCH variables
Δ	Daily adjustment [e.g., in variance]
δ	Coefficients to multivariate regressions variables
ρ	Correlation coefficient between an asset and the market
σ	Standard deviation (σ^2 denotes variance)
Φ	Long-term mean variance

Single letter Latin symbols

E	Expected [e.g., beta]
i	Individual stock or industry
m	market (S&P 500)
N	Number of observations regarding total sample
n	Number of observations regarding a subsample
Q	Ljung-Box test-statistic
t	Trading day
u	Daily return [of a stock or index]
x	Lower limit of event window
y	Upper limit of event window

Other symbols

AAB	Average abnormal beta
AB	Abnormal beta
AR	Abnormal return
B4	Big four [breweries]
CAB	Cumulative abnormal beta shift

CAR	Cumulative abnormal return
CAAB	Cumulative average abnormal beta shift
CAAR	Cumulative average abnormal return
CB	Cross-border M&A
EM	Target in emerging market
MM	Mature market [transactions]
RS	Relative size (transaction volume / market value)
StD	Standard deviation
TV	Transaction volume
Yr	Year

1. Introduction

Corporate finance research frequently concentrates on the issue how decisions of corporate managers and external market events affect shareholder wealth. The maximization of shareholder wealth is one of the central objectives corporate managers aim for, not least because shareholders, as legal owners of the respective corporations, incentive them to do so. Consequently, research on corporate decision making has largely dedicated itself to the broad scope of corporate decisions regarding financing and investment options corporate managers face. Financing comprises the wide set of issues concerning the satisfaction of corporate capital requirements by, e.g., internal financing, bank loans, bond issues, so as primary and secondary equity offerings at the stock markets. Investment decisions focus on how corporate funds are used to ascertain operative business performance and achieve strategic goals. Thus, funds are used to, e.g., perform research and development, acquire machinery and equipment, extend production facilities, expand into foreign markets, or involve into mergers and acquisitions (M&A). In addition to organic growth, M&A transactions enable corporations to, a.o., achieve business expansion at an accelerated pace, diversify a potentially risky portfolio, achieve cost synergies, or obtain access to new investment opportunities.

Apart from internal decision making, shareholder wealth is also affected by external events to the corporation. These external events may impact the economy as a whole but most probably impact certain corporations more than others due to geographical location, industry affiliations, or company specific characteristics like financial leverage, diversification degree, cost structure, or private versus public ownership. Hence, corporate finance scholars have also dedicated a considerable amount of their research effort to scrutinize how expected and unexpected external events impact shareholder wealth. Among expected events one understands, e.g., the scheduled announcements of GDP growth rates, unemployment rates, federal fund rate adjustments at regular meetings, election outcomes, or newly enacted laws. All these may in fact carry some unexpected information, too, as announcements can exceed or fall below expectations. Nonetheless, they clearly distinguish from truly unexpected events such as major corporate or bank failures, natural catastrophes, terrorist attacks, or nuclear power plant disasters like those of Chernobyl or Fukushima.

One econometric method to analyze the effect of events on shareholder wealth that has been very popular among financial economists over the past decades is the event study. Corrado (2011) provides an extensive overview of the refinement of the event study methodology over more than forty years. Since the landmark papers by Ball and Brown (1968) and Fama et al. (1969) this methodology has been one of the central workhorses of financial researchers and also migrated to other disciplines like economics, marketing, and management so as law, history, and political science. The economic skeleton of the financial event study is based on the 'market model' that in itself derives from the capital asset pricing model (CAPM) of Sharpe (1964). To simplify, the respective theory presumes that a stock's daily return depends on the general market return, and the relationship is described by the beta-factor. Beta is also denoted as the systematic risk indicator implying that the more risk a stock carries in relation to the market the higher is its expected return. Under the assumption that beta is stable throughout time, one can determine abnormal returns around the event under investigation subtracting actual returns from expected returns dependent on actual market returns. The statistical testing of the abnormal returns enables the scholars to draw a conclusion on potential shareholder wealth effects induced by the event under examination.

However, the assumption that the beta-factor as systematic risk estimator is stable throughout time has sparked off a wide discussion among financial economists. E.g., Jagannathan and Wang (1996), Collins, Ledolter, and Rayburn (1987), Bos and Newbold (1984), Sunder (1980), and Fabozzi and Francis (1978) have argued against the assumption of beta stability, and have shown that beta is actually stochastic and thus conditional. In such a setting beta may also be determined as the product of an asset-to-market correlation coefficient and the asset's volatility divided by the market's volatility. Engle (1982) and Bollerslev (1986) found that financial stock market data follows certain stylized facts among which volatility clustering over time is highly prevalent. There are time periods of high volatility and others of low volatility showing that the autoregressive characteristic of daily volatility estimates cannot be neglected. Consequently, they developed a model to estimate daily volatility measures incorporating mean reversion, the previous day's volatility measure, as well as the respective stock return. The generalized autoregressive conditional heteroscedasticity (GARCH) model thus enables scholars to estimate daily volatility changes and, as a subsequent step, daily adjustments in the systematic risk estimator beta.

The studies discussed in the main part of this book all concentrate on short-term risk adjustments applying the above elucidated GARCH model. My findings on risk changes add to previous evidence applying event study methodology which in many cases leads to the observation of significant abnormal returns. The new perspective on, e.g., the systematic risk estimator beta uncovers that the assumption of a constant beta over time can result in erroneous statements of shareholder wealth effects. As positive and negative abnormal returns are determined it is common practice to derive conclusive statements on increased shareholder wealth or destruction of the same. However, when the risk analysis returns evidence of an increase or decrease in the beta-factor this critically affects the consolidated effect on shareholders' wealth. Exemplarily, a beta risk increase may offset a positive abnormal return as the basic assumption underlying the latter is a stable systematic risk factor. Any combination of zero, positive, and negative abnormal returns and risk changes may occur which can potentially result in offsetting or enhancing wealth effects. There may of course also be no interference.

In the five consecutive studies this dissertation comprises I address the implications on shareholder wealth of two strategic investment decisions of corporate managers and of two external events. The respective research titles are:

I. Does Vertical Diversification Create Superior Value? Evidence from the Construction Industry

II. How Consolidation Changes the Risk Profile of a Whole Sector: Evidence from the Brewing Industry

III. Bank Failures and Real Estate Investment Trust (REIT) Returns

IV. Bank Bailouts: REITs and their Performance as Financial Stocks

V. Banking Crisis and Government Intervention: How Bailouts versus Failures Affect Financial Stocks

The first two studies concentrate on systematic risk changes and build upon previous evidence that has yet identified abnormal returns. The first study sheds light on the decision making process corporate managers of construction companies face when they pursue a M&A transaction which urges them to choose between horizontal (predominant business), vertical (related business), or lateral (unrelated business) diversification. The second study sheds light on the strategic investment decision brewing managers face in a contracting market and a strongly consolidated industry where M&A target companies are scarce and competition for these is fierce.

The other three studies concentrate on external events induced by the recent financial crisis starting in 2007, measuring the effect of U.S. bank failures and bank bailouts on REITs and other financial stocks. Study three examines a set of 130 bank failure dates and their impact on REIT mean returns and beta risk to potentially uncover whether REITs as hard assets outperform common equities during times of high market uncertainty. Study four shifts focus to the opposite event, being bank bailouts, and examines if moral hazard theory is supported by positive systematic risk changes and REITs as hard assets here devalue relative to other equities. Finally, study five builds upon the reverse findings of studies three and four and extends the scope to the general financial services industry. Incorporating a second major stylized fact of financial return data, the asymmetry in volatility reaction to positive and negative return shocks, the exponential GARCH (EGARCH) is applied to even better observe the opposing effects of bailouts and failures on stock return volatility and beta.

In the subsequent paragraphs the reader will be provided with some more detail on the chapters of this dissertation, elucidating the background, relating the respective research questions to previous evidence, reporting core findings, and commenting on general implications.

Study I, chapter 2 – Corporate Diversification Strategies

The discussion of diversification discounts is one of the most controversial in corporate finance and strategic management. I am eager to reexamine this issue from the standpoint of vertical versus lateral diversification, and horizontal growth through construction industry M&As. I build on previous evidence of positive acquirer abnormal returns for vertical M&A, and I add new insight into stock return risk. Considering the high idiosyncratic risk levels of builders, I expect to find considerable informational content in systematic risk (beta) behavior, which has been neglected to date. In fact, I find that vertical M&A experience a negative asset beta shift, lateral M&A experience an increase in systematic risk, and only horizontal M&A exhibit no risk changes. Hence, risk and return-induced wealth creation from vertical M&A shows that related industrial diversification (vertical M&A) is superior to unrelated (lateral M&A) – at least in the construction industry.

Study II, chapter 3 – Consolidation and Rival effects

Consolidation has been and continues to be a major trend in the brewing industry as multi-national breweries, in an attempt to offset the decline in beer volumes in mature markets (in particular Western Europe), seek to expand their activities into new emerging markets. I analyze short-term risk implications of horizontal M&A on acquirers and rivals in the brewing industry. Based on a sample of 75 takeover announcements between 1998 and 2010, I document significant negative acquirer abnormal risk shifts for cross-border transactions, especially when targets are in emerging markets. Furthermore, I document evidence of significant abnormal negative risk shifts to the "big four" (Anheuser-Busch Inbev, SABMiller, Heineken, and Carlsberg) when announcing a M&A and significant positive abnormal risk shifts when missing out on a potential

M&A as rivals. Thus, my study shows that consolidation may result in contrary risk implications on different industry players.

Study III, chapter 4 – Bank Failures and REITs

I reexamine the case whether Real Estate Investment Trusts (REITs) behave rather comparable to common equities versus to real estate. Bredin, O'Reilly, and Stevenson (2007) analyze monetary rate increases, as external shocks, increasing the price of liquidity. They find decreasing REIT returns and increasing risk levels similar to common equities. I introduce bank failures, which can be contagious, as an internal shock that may decrease liquidity supply. My results show that Equity REITs in general and especially Retail, Residential, and Healthcare REITs experience positive abnormal returns relative to common equities. This implies that market participants regard Equity REITs comparable to direct real estate which as "hard asset" is seen as "safe haven" during turbulent times. In contrast, indirect real estate investments like Mortgage REITs devalue. I show that primarily the capital investment type but also property focus are major determinants whether REITs behave like real estate versus common equities during crisis.

Study IV, chapter 5 – Bank Bailouts and REITs

Previous evidence on monetary policy finds positive value effects on REITs when federal fund rates are decreased in order to strengthen the market's liquidity pool. Bank bailouts can also be seen as positive monetary shocks to capital markets as government intervention restores bank liquidity and by this market liquidity. However, empirical evidence on moral hazard shows that bailing out banks may weaken risk adversity and potentially benefit relatively risky assets. Therefore, I expect REITs as hard assets to be outperformed by the market on a short-term scope. In fact, I find supporting evidence, i.e., negative abnormal returns and beta risk increases for Equity REITs around the bank bailout announcements. This extends my previous research on bank failures and REIT returns where positive abnormal returns around bank failures prescind Equity REITs as a safe haven in times of market uncertainty. Furthermore, I complete my research by investigating other financial stocks, finding comparable results.

Study V, chapter 6 – Banking Crisis and Government Intervention

After the devastating bankruptcy of Lehman Brothers the U.S. government announced a bailout program of unprecedented size to prevent a cascade of bank failures and by this a potential meltdown of the total system. The intention was to calm down financial markets preventing a spillover effect on the real economy. I analyze the effects of bank bailouts, controlling for failures, on the financial services industry (FSI) to scrutinize if government intervention properly calmed markets without benefitting financial stocks. I find FSI subsectors react ambiguously to bank bailouts: while banks show two-sided significant abnormal returns, insurances react positively and REITs negatively. Reverse effects are found for bank failures. EGARCH (1,1) estimated variances generally decrease around bailouts indicating enhanced system stability. However, beta strictly increases around government intervention and offsets any respective positive FSI valuation effects. Policy makers should be aware that short-term stability comes at the price of increased systematic risk.

I draw three main implications of my research that are relevant to corporate managers so as corporate and private investors. First, optimal diversification at the firm level is different from that of individual (private) investors. Corporate managers should possess exceptional expertise in their area of business and I find it to be valuable to limit diversification efforts to related businesses. Unrelated or wider diversification should only be done by the uninformed investor who is commonly the private investor optimally holding a mixed portfolio of (all) stocks and risk free securities.

Second, my research shows that corporate managers face rigorous capital markets when their companies find themselves in the highly competitive environment of a declining market. Missing out on potential growth opportunities can result in a harmful two-sided worsening of their stocks' risk-return profile. To ensure the survival end enforce the competitive advantages of their companies corporate managers must excel in their efforts to realize growth opportunities – either via horizontal M&A or, if missing out, via an adjustment of their corporate strategy. Otherwise, they will destroy shareholder wealth relative to their rivals.

Third, my analyses on REITs show that they should not unambiguously be seen as hard asset and save haven investment in times of high market uncertainty. In fact REITs can generally behave very much like common stocks and less like conventional real estate especially during times of crises. Nevertheless, Equity REITs and among this type low risk Residential and Healthcare REITs clearly outperform common equities around bank failures that increase market uncertainty. As real estate is core to private investments and most businesses it is vital to understand that REITs may differentiate substantially from pure real estate values and only certain REIT types carry the hard asset and save haven properties.

2. Does Vertical Diversification Create Superior Value? Evidence from the Construction Industry[1]

2.1. Introduction

The literature on corporate finance and strategic management has explored the motives of corporations to pursue horizontal and diversifying merger and acquisition (M&A) activity extensively, and many studies have analyzed their respective valuation effects through the capital markets. Nevertheless, the discussion of whether diversification ultimately creates or destroys acquirer shareholder wealth is ongoing.

Martin and Sayrak (2003) identify three main strains of perspectives on shareholder value impact through diversification. The first posits that diversification is value-destroying. For example, Rajan, Servaes, and Zingales (2000) show how agency problems can cause distorted investment allocations, which can lead to underperformance of the total entity. Thus, diversified firms are thought to trade at a discount relative to comparable stand-alone companies. Stiroh and Rumble (2006) provide further evidence of the dark side of revenue diversification in the financial sector.

The second strain argues that corporate diversification does not destroy shareholder value, and believes previous evidence has neglected the endogeneity of diversification. An example of this line of reasoning is Rawley (2010), who finds that the economies of scope achieved through diversification easily offset any increased costs of coordination and organizational rigidity.[2]

The third strain finds that significant premiums for diversified firms could be identified by, e.g., matching samples of single-business firms. Elsas, Hackethal, and Holzhäuser (2010) also find positive evidence for the economies of scope explanation by studying bank diversification. Better access to capital markets may be another reason for the enhanced value of diversified firms, as noted by Hadlock, Ryngaert, and Thomas (2001).

---

[1] I thank Dirk Schiereck and Andreas Pfnür for valuable comments and insights. I am also grateful for the feedback of conference participants at the annual meetings of the Academy of Economics and Finance 2010 in Houston and of the American Real Estate Society 2011 in Seattle.

[2] Matsusaka (2001), Goold and Luchs (1993), and Davis and Duhaime (1992) provide further evidence of diversification and firm performance in the management literature. The latter concentrates on vertical integration.

We are eager to reexamine the issue of shareholder wealth implications through diversification. We choose corporate M&A as our focus in order to analyze the value implications of diversifying (vertical versus lateral), and of horizontal external growth strategies. This article contributes to recent evidence put forth by Dos Santos, Errunza, and Miller (2008) that unrelated M&A are value-destroying.[3] We control for industry effects, and examine the research question from a case study viewpoint. We expect to obtain a clearer view concentrating on one fast-changing and consolidating industry, and we hope to extrapolate our potential findings to other industries at a later stage.

We believe the construction industry is an optimal example for this research for three reasons: 1) It is a highly fragmented industry that is also undergoing a great deal of consolidation, where all three strategic directions of M&A (horizontal, vertical, and lateral) can be observed (see, e.g., Green, Fernie, and Weller, 2005); 2) it is exposed to an exceptionally high imminent risk level because of its players' concentrated portfolios, its highly volatile construction services markets, and the inherent risks of large-scale commercial and infrastructure projects; and 3) there is not yet a considerable bulk of empirical evidence on construction industry M&A that applies event study methodology or even performs risk analysis.

Rottke, Schiereck, and Pauser (2011) and Choi and Russel (2004) find the strongest positive abnormal returns for acquirers that diversify vertically, followed by horizontally. The value implications of lateral M&A in the construction industry are rather negative, indicating that the benefits of diversification are probably exceeded by the higher monitoring and post-merger integration costs.

However, previous studies have lacked evidence on true risk dynamics. Mansi and Reeb (2002), for example, highlight the importance of the risk effects of diversification, and they find that discounts stem from risk-reducing effects. Nevertheless, they proxy for risk with a market analysis and the book values of debt and equity. Furthermore, financial economists applying event study methodology appear to have neglected the observation of risk behavior as the second moment of stock returns, next to the mean as the first moment. They assume a stable beta risk through time. But this assumption is questionable; we expect to find considerable information from beta (systematic risk) behavior as a core determinant of shareholder wealth creation.

---

[3] See Moeller and Schlingemann (2005) for a discussion of lower announcement returns of global and industrial diversification.

Our objective is to highlight the importance of second moment returns relative to first moment returns. We argue that value effects around M&A announcements may have historically been under- or overestimated, because potential risk shifts parallel to observed mean shifts in stock returns were neglected. We anticipate that diversification will act as a risk determinant, especially in an industry such as construction with a high imminent risk level. Transaction direction primarily affects portfolio characteristics. The degree of diversification along correlated cash flows (horizontal M&A) increases less than it would along uncorrelated cash flows. But managerial efficiencies may be more difficult to translate into unrelated businesses (lateral M&A) than into an extension of the already predominant processes (vertical M&A).

We aim to contribute to the literature in three ways: 1) We analyze short-term systematic risk adjustments around M&A announcements in the construction industry. We contrast risk and return behavior by reexamining potential abnormal returns, thus ensuring comparability to previous evidence; 2) we analyze long-term risk behavior. This allows us to shed light on unlevered beta behavior, which reflects pure business risk, separate from financing risk; and 3) we perform several multivariate regressions on risk dynamics to obtain further evidence on the potential determinants. Overall, we expect to observe significant beta risk adjustments that will support our argument of how diversification affects acquirer risk levels, and add economic insight to the observed abnormal returns.

Our sample consists of 119 international M&A transactions that involve construction industry acquirers. The sample is comprised of sixty-one horizontal, thirty-two vertical, and twenty-six lateral transactions. The M&A announcement dates cover 1988 through 2007; approximately 50% of the transactions take place in 2003 or later. We apply an adapted beta risk event study based on a GARCH (1,1) model in order to measure abnormal systematic risk changes in equivalent windows as abnormal returns. We use standard event study methodology to verify previous observations for our data sample; we then examine long-term asset beta behavior and equity beta behavior (the former is applicable on only an annual basis because of the limitations of financial statement data). By performing several multivariate regressions, we aim to obtain further evidence for our short- and long-term risk analysis results.

For our total sample, we find highly significant positive cumulative average abnormal returns *(CAARs)* for event windows [-1,1] and [0,1], and another six positive *CAARs* that are less significant. However, we do not observe any abnormal beta changes for the total sample. All three transaction directions – horizontal, vertical, and lateral – lead to positive abnormal returns. The respective short-term abnormal beta changes exhibit positive means and medians for horizontal and lateral M&A. Two of the latter are significant. Negative abnormal beta shifts for vertical M&A are not statistically significant. However, we find evidence that builders who vertically acquire assets can experience beta decreases even up to the second post-M&A year. Our multivariate regressions support these findings to a considerable extent, further indicating a long-term increase in equity beta following lateral M&A.

In summary, our empirical analyses illustrate that beta risk observation around M&A in the construction industry can add new insights into acquirer wealth effects. Previous evidence on abnormal returns underestimates the two-sided value-creating risk-return profile shift for vertical M&A. Lateral M&A lead to both short- and long-term increases in beta risk, which most likely result in a decrease in shareholder value. But for horizontal M&A, no significant risk adjustments are found, which adds to the ambiguously positive view of the return shifts found in previous studies. Hence, risk and return-induced wealth creation from vertical M&A shows that related industrial diversification is superior to unrelated, at least for the construction industry.

The remainder of the paper is structured as follows. Section 2.2 provides a literature review, while section 2.3 explains our data selection and empirical methodology. Section 2.4 presents our results, and section 2.5 concludes and offers ways to structure future research.

2.2. Literature Review

The intuition behind our literature review is threefold: First, we argue why M&A represents a good case for analyzing the effect of diversification on shareholder wealth. We discuss M&A motives along horizontal, vertical, and lateral transaction lines, and we present empirical evidence on the risk adjustments due to diversification. Second, we discuss the consolidation of the construction industry, and argue why it provides a singular case for observing risk dynamics. Third, we derive case-specific hypotheses for how the risk and return behavior of our total sample and the three transaction directions can strategically drive shareholder wealth.

2.2.1. M&A Motives across Transaction Directions

M&A as part of corporate strategy has an immediate effect on the degree of acquirer diversification. As such, it is unmatched in either impact or measurement precision by comparable organic growth initiatives. M&A thus fulfills two critical requirements for a well-defined financial econometric analysis: 1) it constitutes a homogeneous strategic business issue in corporate finance, and 2) the date new information becomes public is presumably precisely provided (announcement date).

The motives for vertical or lateral M&A transactions are built on risk reduction through (related or unrelated) industrial diversification. At first glance, for example, a product market diversification may seem to result in an increased number of uncorrelated cash flows. The likelihood of a single business issue putting the firm as a whole at risk is decreased. And several financial economists support the notion of reduced cash flow variability from a diversification strategy that is similar to the way investors build efficient portfolios (e.g., Amit and Wernerfelt, 1990; Amit and Livnat, 1988a, 1988b).

However, contrary evidence suggests that unrelated diversification increases firm risk (Lubatkin and Chatterjee, 1994; Lubatkin and O'Neill, 1987). The managerial difficulties caused by added firm complexity may outweigh any of the risk reduction effects achieved through the diversification. Using the investor view as an analogy, Scharfstein (1998) and Amihud and Lev (1981) find that diversification strategies based on simple risk reduction motives have often proven to be value-diminishing, because stockholders can most efficiently reduce risk themselves by holding diversified portfolios.

Agency theory further suggests that the lateral diversification resulting from the separation of ownership and control can affect managers' incentives negatively. As Jensen (2005)[4] and Jensen and Meckling (1976) have noted, the separation of ownership and control may lead to agency costs if management pursues unrelated mergers that are not being closely monitored by, e.g., large blockholders. Managers who diversify in order to achieve reduced employment risk may end up obtaining private benefits at the expense of shareholders.

Wang and Reuer (2006) measure the potential risk reduction of M&A. As their risk measure, they use the correlation in stock returns between the acquirer and the target for the three-year period prior to the transaction. They thus obtain an indication of how the merger affects both firms' future conjoined risk levels, but they do not actually monitor acquirer risk behavior, as we wish to. Also, because Wang and Reuer (2006) focus on so-called firm-specific investments, they find that the risk reduction potential of M&A transactions is positively correlated with a firm's dependence on firm-specific investments.[5]

---

[4] A high level of firm valuation increases managerial discretion even further, and may thus lead managers to pursue poor acquisitions when they have run out of good ones.

[5] Firm-specific investment takes into account firm stakeholders, but specifically excludes financial share- and debtholders. Firm stakeholders are those who have the greatest specific knowledge of the firm and its administrative processes, or customers and suppliers. Those

Montgomery and Singh (1984) investigate the relationship between systematic risk and diversification strategy for ninety-nine firms whose betas are based on a sixty-month observation period. Rather than single transactions, they concentrate on a company's general diversification strategy over a longer horizon. Their findings constitute significant differences in mean beta among horizontal (single business, dominant), vertical (related constrained, related linked), and lateral (unrelated portfolio) diversification strategies. Lateral diversification strategies lead to higher betas. In addition, Montgomery and Singh (1984) find that conglomerates actually experience more volatility than the market itself.

2.2.2. Consolidation of the Construction Industry

We note that the construction industry is a global but still highly fragmented industry. As Green, Fernie, and Weller (2005) state, builders operate primarily in local markets, with a few large enterprises that are able to compete for the most complex worldwide projects. As a result, average firm size remains relatively low compared to the competitive landscape of, say, financial services, logistics, or automotive production.

Moreover, builders currently face little regulation, compared again with, say, financial services providers or pharmaceutical producers. The markets for construction services feature a considerable degree of imminent risk, however. Housing demand shows high growth at times, but it is also extremely susceptible to risk. E.g. Ellis (2010) provides respective evidence with his in-depth analysis of the boom-bust cycle in the U.S. housing market. Furthermore, large commercial real estate or infrastructure projects alternately add bulk and cluster risks to builder portfolios. Consequently, we characterize construction companies on average as relatively small, with fairly concentrated portfolios that are vulnerable to local economic problems, often possess less efficient organizational structures and risk management, and are highly dependent on volatile or clustered markets.

In short, builders carry a high level of idiosyncratic risk. Analyzing risk as the second moment of stock returns should reveal valuable insights. Concentrating further on M&A should increase the probability of seeing risk shifts, because builders have increasingly pursued M&A strategies to overcome the inherent vulnerability of their small businesses.

Cuervo and Pheng (2005) provide evidence for the internationalization of Singaporean construction corporations. They find that reputation and brand

---

assets can be potentially costly for the stakeholders. Thus, stakeholder willingness to increase firm-specific investments is inversely correlated with a firm's total risk behavior.

quality are highly relevant factors. Delany and Wamuziri (2004) argue that plain horizontal mergers in the construction industry build on increased reputation through size by signaling higher technical competencies and a more solid capital base that allows for more advanced financing of large-scale projects. Cross-border horizontal transactions can also reduce the impact of national economic downturns.

Flanagan and Norman (1993) state that construction firms tend to be dependent on the success of isolated large-scale projects that have limited time frames. In an effort to reduce this high idiosyncratic risk level, they aim to extend the value chain through vertical acquisitions. As an example, real estate facilities management is based on long-term contracts with continuous income streams, and thus has more moderate risk structures at lower absolute levels.

Recent evidence on M&A in the construction industry by Rottke, Schiereck, and Pauser (2011) analyzes 106 transactions from 24 countries between 1986 and 2006. The authors concentrate on M&A-induced return-driven value effects, and find evidence for acquirer *CAARs*. Total sample announcement effects show positive *CAAR* medians at the 1% significance level, and insignificant *CAAR* means. For the transaction direction subsamples, the results hold for only thirty vertical M&A transactions. Acquirer medians are positive, while means remain insignificant. For a small subsample of fourteen lateral M&A transactions, Rottke, Schiereck, and Pauser (2011) find a negative *CAAR* median with low significance. A multivariate regression analysis confirms the results of the positive return effects through vertical M&A, and the negative return effects through lateral M&A.

Previous research by Choi and Russel (2004) analyzed 171 U.S. transactions between 1980 and 2002. They found significantly positive acquirer *CAAR* mean and median values around the announcement date. For a conglomerate subsample of thirty-eight transactions, they found insignificant means and medians. Horizontal M&As are represented by the "geographic expansion" subsample within the business relatedness (strategic objectives) analysis. The sample is comprised of seventy-two transactions with positive abnormal returns. Finally, the vertical M&A subsample (n = 36) exhibits highly significant positive means. However, a multivariate OLS regression cannot confirm the positive impact of the vertical transaction direction.

2.2.3. Hypotheses for Risk and Return Behavior

Previous evidence on the first moment behavior of stock returns is quite similar, which makes it easier for us to derive hypotheses for our sample. The second moment of stock returns illustrates a certain risk effect through diversification. Nevertheless, the evidence is ambiguous, and the hypotheses for our sample of construction industry M&A are more difficult to derive. A lack of risk analyses around individual M&A events using a comparable methodology of consecutive beta estimates before and after transactions, as well as the diversity of transaction directions, makes it difficult for us to derive case-specific hypotheses.

For our total sample risk and return analysis, we expect a positive return shift that will unambiguously confirm previous evidence. For systematic risk, we expect to find a negative trend. We regard the diversification effect of vertical and lateral M&A on our total sample as a strong risk absorber. However, we also remain skeptical about the contrary risk implications of lateral M&A on our total sample, which result from the increased uncertainty about monitoring and post-merger integration success.

For horizontal M&A, previous evidence on return-driven value shifts has been ambiguous. We thus expect insignificant (positive) abnormal returns. But for systematic risk, we do not anticipate any significant adjustments, in line with market theory. We generally expect that M&A within an industry should result in combined entities with risk characteristics that are similar to the acquirers.

From previous evidence on vertical M&A transactions, we expect to obtain highly significant positive abnormal returns. This value-increasing return shift should be tied to a decrease in beta risk as builders expand into construction-related businesses and real estate services. These businesses tend to carry less imminent risk, and can be substantial sources of business diversification. The decrease in risk should be even stronger for asset beta, because it reflects pure business risk separately from financing risk.

Event study evidence on lateral M&A shows a tendency toward negative abnormal returns, from which we expect at least some negative return shifts at low significance. For risk adjustments, we should observe an increase in equity beta, because previous evidence on conglomerates has identified higher risk levels. We believe that shareholder skepticism about the probability of success for lateral M&A will outweigh the immediate diversification measures, and thus positively affect asset beta.

Although our primary focus is on the distinctive effects of the three transaction directions on acquirer risk-return profiles, we also explore the impact of transaction size measures and potential biases such as time period and geographic deal characteristics. By analyzing the impact of transaction volume and relative transaction size to acquirer market value on the acquiring builders' risk-return profile, we expect to simply observe a magnifying effect of absolutely and relatively larger M&A. In other words, we expect diversification effects will be enhanced by larger transaction size, but that parallel monitoring and post-merger integration efforts will also increase. We hope to observe no significant effects from time period or geographic characteristics of M&A as potential biases to risk and return behavior.

2.3. Data and Methodology

2.3.1. Data Selection

Our sample of M&A transactions comes from the Thomson One SDC Platinum Database. We consider worldwide transactions with deal values of at least U.S. Dollar (USD) 10 million, announced between January 1, 1988, and December 31, 2007, where over 50% of the bidding firm was acquired. In addition, we only examine mergers where the acquirers are strictly classified as construction businesses and are classified within the 1,500-1,799 SIC code (the construction industry). We further restrict our sample to acquirers that have not pursued an acquisition during the one-year period surrounding the announcement date, where total return index data (RI) shows continuous trading activity[6] from two years prior to two years after the event. We also exclude any acquirers classified as penny stocks within that time period, and we only consider countries where local market indices are available. Our return data for the acquiring construction firms and for the home indices come from Thomson Financial Datastream.

Our final sample consists of 119 international transactions from 25 countries, with 13 identified as cross-continental based on a regional clustering within Europe, Asia, North America, and various other single countries. Table A2.1 in the appendix depicts all transactions along the time span of the study, and indicates that M&A activity (in the construction industry) follows economic cycles. Table A2.2 shows acquirers and targets by country.

The reasoning for using local market indices over global (industry) or local industry indices is twofold: First, we prefer to control for cross-regional imbalances in economic cycles, which eliminates a global index. Applying contrary methodology could result in beta biases from regional aspects. Second, using a (local) industry index we would measure the systematic risk (beta) of an index composed significantly of our sample firms. This is a critical step, because firms under analysis could impact index performance significantly and lead to endogeneity problems. We cannot rule out any endogeneity from using local market indices, but it should be negligible compared to local industry indices.

---

[6] Infrequent trading activity is defined as by Corrado and Truong (2008) with 25% or more of trading days with zero daily returns.

2.3.2. Methodology

Our empirical methodology is twofold. In our primary focus are the short- and long-term analyses of systematic risk behavior, but we also perform a standard event study. We can thus ensure a high level of comparability between our sample and results and previous M&A evidence from the construction industry.

To study short-term systematic risk (beta) behavior, we use an advanced event study methodology that can measure short-term shifts in beta that exceed or fall short of expectations. The mathematical intuition is equivalent to the standard event study. Our risk event study is comprised of three basic steps: 1) measuring daily acquirer betas, 2) estimating expected betas before M&A announcements, and 3) subtracting measured betas from expected betas to obtain abnormal beta shifts, *AB*. By combining the *ABs* within event windows and determining the arithmetic mean across events, we obtain the cumulative average abnormal beta shifts, *CAABs*. As scholars of financial economics will note, measuring time-variant beta is a breach of the basic CAPM assumption that beta is stable over time. However, financial economists such as Jagannathan and Wang (1996), Collins, Ledolter, and Rayburn (1987), Bos and Newbold (1984), Sun-der (1980), and Fabozzi and Francis (1978) have argued against the assumption of beta stability, and have shown that beta is actually stochastic and thus conditional.

To determine time-varying variance and beta measures, we apply the generalized autoregressive conditional heteroscedasticity (GARCH) method, one of the most commonly used techniques to observe changes in risk level. This methodology was first introduced by Engle (1982), and refined by Bollerslev (1986). It was then further advanced to a multivariate GARCH (M-GARCH) model by Bollerslev (1990). See Giannopoulos (1995) for an empirical implementation of GARCH models, where he finds that such volatility modeling techniques can also be used to measure systematic and unsystematic components.

Brooks, Faff, and McKenzie (1998) investigate the forecasting accuracy of three techniques for obtaining time-variant beta: the M-GARCH, a time-varying beta market model suggested by Schwert and Seguin (1990), and the Kalman filter for a sample of returns on Australian industry portfolios. Faff, Hillier, and Hillier (2000) also apply an M-GARCH approach to model time-varying beta in their study of alternative beta risk modeling techniques. They add the assumption of a constant correlation coefficient between the returns of an asset and the respective market returns. The behavior of this correlation coefficient is a key component of our risk event study that determines *CAABs*.

We use a simple GARCH (1,1) model as shown in Equation (2.1), which defines a specific asset's time-varying variance, σ_{it}^2, as the weighted sum of the long-term average variance rate, Φ_i, the previous day's squared return, $u_{i,t-1}^2$, and the previous day's variance, $\sigma_{i,t-1}^2$. The weighting factors γ_1, γ_2, and γ_3 are determined by applying the iterative maximum likelihood (ML) technique as discussed in Hull's (2006).[7] Equation (2.1) shows the GARCH (1,1) model:

$$\sigma_{it}^2 = \gamma_1 \Phi_i + \gamma_2 u_{i,t-1}^2 + \gamma_3 \sigma_{i,t-1}^2 \tag{2.1}$$

Table 2.1 provides an overview of our GARCH estimations via ML. To illustrate our total sample, we display three GARCH specifications for the 119 builders and three specifications for the respective sample of market indices. We use the weight of persistence in volatility, γ_3, rather than the weight of shock vulnerability, γ_2, or the weight of long-term variance, γ_1, to identify representative datasets for our diverse sample.

To validate our GARCH calibration, we apply the Ljung-Box test for serial correlation (see Ljung and Box, 1978), and the ARCH test for heteroscedasticity in stock returns (Engle, 1982). The test statistics show that GARCH can significantly decrease serial correlation and heteroscedasticity in squared stock returns, u^2, when stock returns are standardized by the GARCH estimated variance, u^2/σ^2. As per prior literature, we choose to test a short-term lag structure of one to three days, and medium-term lag structures of five, ten, and fifteen days. Note that our test statistics show that neither heteroscedasticity nor serial correlation are continuously detected for squared returns, nor are they fully eliminated for standardized squared returns. But we believe the GARCH (1,1) model is sufficient to ensure a well-specified risk analysis.

---

[7] Further methodological reviews on maximum likelihood can be found in, e.g., Wooldridge (1990), Greene (2004), and Aït-Sahalia and Kimmel (2007).

Table 2.1: GARCH Calibration and Tests: Sample Overview

Selection by variance persistence (γ_3)			Builders			Markets		
		relative weight of:	Min	Max	Median	Min	Max	Median
GARCH	γ_1	long-term variance	58.9%	0.9%	4.4%	28.6%	0.8%	1.4%
	γ_2	moment of shock	22.7%	1.7%	8.5%	14.2%	5.6%	8.6%
	γ_3	variance persistence	18.4%	97.4%	87.1%	57.2%	93.6%	90.0%
Ljung-Box Test	u^2	*lags 1/2/3*	1/1/1	1/1/1	1/1/1	1/1/1	1/1/1	0/1/1
		lags 5/10/15	1/1/1	1/1/1	1/1/1	1/1/1	1/1/1	1/1/1
	u^2/σ^2	*lags 1/2/3*	0/0/0	1/1/1	0/0/0	1/1/1	0/0/0	0/0/0
		lags 5/10/15	0/0/0	1/1/1	0/0/0	1/1/0	0/0/0	0/0/0
ARCH Test	u^2	*lags 1/2/3*	1/1/1	1/1/1	0/0/0	1/1/1	0/0/0	0/1/1
		lags 5/10/15	1/1/1	1/1/1	0/0/0	1/0/0	1/1/1	1/1/1
	u^2/σ^2	*lags 1/2/3*	0/0/0	1/0/0	0/0/0	0/0/0	0/0/0	0/0/0
		lags 5/10/15	0/0/0	0/0/0	0/0/0	0/0/0	0/0/0	0/0/0

The estimation of the GARCH (1,1) model. The Ljung-Box (1978) test for serial correlation and the ARCH test for heteroscedasticity of squared stock returns u^2 versus GARCH-standardized squared stock returns u^2/σ^2: For each lag 1, 2, or 3, and 5, 10, or 15, zero indicates no serial correlation or heteroscedasticity, and one indicates the opposite at a 5% significance level.

Our next step is to obtain beta measures. We apply the M-GARCH model that observes actual betas as the product of a constant correlation coefficient, ρ_{im}, between the specific acquirer returns and the respective market returns, and the time-varying acquirer standard deviation, σ_{it}, divided by the time-varying market standard deviation, σ_{mt}.

$$\beta_{it} = \frac{\rho_{im}\sigma_{it}}{\sigma_{mt}} \tag{2.2}$$

When we complete this empirical groundwork, we have the following data on hand: daily variance measures for all construction industry acquirers and respective market indices, daily correlation coefficients for all acquirers, with their corresponding market indices based on a one-year lag structure, and daily beta measures for all acquirers.

For the risk event study, we use the correlation coefficient thirty days prior to the event to obtain systematic risk estimates, $E(\beta_{it})$ (this is equivalent to the use of beta in standard return event study). When confronted with event-induced variance, it is critical to analyze moments of stock returns closely around the event date. In the case of event studies, event-induced variance is best accounted for by applying parametric test statistics as per Harrington and Shrider (2007). These statistics use standard errors that are robust to cross-sectional variation in true abnormal returns. For our risk event study, we must separate event-induced variance from the event's persistent effect on beta risk, e.g., on the asset-to-market correlation coefficient. Thus, we decide to keep the acquirer stock and market index return standard deviations time-variant (no forecasting), just as the market return does in the standard event study. We can then determine that the abnormal beta shift, *AB*, is the difference between the expected and the measured beta from ten days prior to ten days after the event.

$$AB_{it} = E(\beta_{it}) - \beta_{it} = \frac{\rho_{im}\sigma_{it}}{\sigma_{mt}} - \beta_{it} \tag{2.3}$$

By combining the *ABs* along the differently sized event windows, we obtain cumulative abnormal beta shifts (*CAB*s). By applying the arithmetic mean of the *CABs* across the number of events per analysis, *n*, we can then obtain the *CAAB*s. We then use, as per Massey (1951), the Kolmogorov-Smirnov test to determine whether *CAABs* are based on normally distributed data. *CAAB* means and medians are also tested for differences to zero with the standard parametric t-test and the non-parametric Wilcoxon signed rank test. We refrain from bootstrapping t-statistics of our risk event study as we will do in the case of the subsequently discussed standard event study. In a test scenario bootstrapping returns p-values indicating highly significant t-values at unconventional levels which evokes a credibility issue. Due to the fact that previous research lacks evidence and recommendations on bootstrapping around beta risk changes we decide not to apply this method.

For the return, we use a standard event study methodology, following MacKinlay (1997). To summarize, this approach uses a market model similar to the Capital Asset Pricing Model (CAPM) of Sharpe (1964) and Lintner (1965). By performing simple OLS regressions from 280 to 30 days prior to the event, we can obtain case-specific betas as systematic risk estimators, from which we can forecast the expected acquirer returns dependent on actual market returns. We then subtract the actual observed acquirer returns around the event dates from the expected returns, and obtain the *AR*s. On a single event basis, the cumulative abnormal returns *(CARs)* would be combined along the differently sized event windows. And by taking the equally weighted averages of the *CARs* with respect to the number of events, we obtain the *CAAR*s. The Kolmogorov-Smirnov test is used to verify whether these *CAARs* are based on normally distributed *CARs*. Finally, we test the *CAAR* means for differences from zero using the Boehmer, Musumeci, and Poulsen's (1991) adjusted t-test (Boehmer test) and the bootstrapped[8] and skewness adjusted parametric t-test similar to Lyon,

---

[8] Based in each case on 1,000 random subsamples of mean-adjusted CARs with sample size n/2 for which skewness adjusted t-statistics are determined.

Barber, and Tsai (1999) and Kramer (2001). Medians are again tested using the non-parametric Wilcoxon signed rank test.

2.4. Results

2.4.1. Structure of the Results' Section

Our results section begins with an analysis of our total sample, presenting both the risk and return perspectives. We discuss our findings from the standpoint of previous evidence and our hypotheses. We then perform various subsample and multivariate analyses to identify the relevant determinants and potential biases that drive our results.

To determine the effect of potential determinants on return and risk behavior, we measure the impact of *transaction direction*, *transaction volume*, and the *relative size of transaction volume to acquirer market value*. Moeller, Schlingemann, and Stulz (2004) find comparable evidence that firm and transaction size are major determinants for a significant decrease in shareholder value around M&A transactions. Regarding any potential biases, we control for selected time periods to reflect individual economic cycles and geographic regions.

We also perform a long-term equity versus asset beta analysis. As mentioned earlier, to determine the outline of this analysis, we obtained company-specific financial leverage ratios from annual Thomson One Banker data. Again, transaction direction, size, time, and geographic characteristics are all examined to determine their impact.

2.4.2. Short-term Risk and Return Behavior

We begin our examination with a total sample analysis, so that we can measure the effect of the construction industry-related M&A from our sample on acquirers' first moments of return (i.e., means). As Table 2.2 shows, we confirm previous evidence of positive abnormal returns in eight of our eleven short-term event windows.

As the results of the Kolmogorov-Smirnov test show, we cannot assume a normal distribution of *CARs* for the [-2;2] to [0;10] event windows at a 5% significance level. Thus, the Wilcoxon signed rank test as a non-parametric test statistic is the appropriate measure to use for determining significance throughout these event windows. However, the presumably normally distributed CARs of the event day and the two days surrounding it support the observation of a positive return effect induced by construction industry M&A. This observation is further validated by the Boehmer test and the bootstrapped t-statistic.

Table 2.2: Standard Event Study: All Construction Industry M&A

CAAR [-x; +y]	Mean	Median		KS-test	p	Wx-test	p	Z-stat.	p	t-test	p	N
0 0	0.96%	0.57%	***	0.14	0.02	-2.66	0.01	2.06	0.04	2.14	0.01	119
1 1	1.72%	0.82%	***	0.12	0.05	-3.29	0.00	3.07	0.00	2.98	0.01	119
2 2	1.53%	0.09%	*	0.11	0.08	-1.63	0.10	2.24	0.02	2.24	0.01	119
4 4	1.76%	0.54%	*	0.11	0.10	-1.74	0.08	2.39	0.02	2.49	0.01	119
7 7	1.12%	2.00%		0.12	0.08	-1.21	0.23	1.33	0.19	1.24	0.10	119
10 10	0.62%	-0.07%		0.11	0.11	-0.33	0.74	0.82	0.41	0.61	0.50	119
0 1	1.65%	0.81%	***	0.12	0.07	-3.53	0.00	3.23	0.00	3.16	0.01	119
0 2	1.64%	0.12%	**	0.10	0.14	-2.27	0.02	2.88	0.00	2.89	0.01	119
0 4	1.51%	0.34%	**	0.09	0.26	-2.27	0.02	2.58	0.01	2.60	0.01	119
0 7	1.04%	0.98%	*	0.09	0.27	-1.88	0.06	1.66	0.10	1.48	0.10	119
0 10	0.31%	0.64%		0.12	0.06	-0.56	0.58	0.63	0.53	0.42	0.50	119

***, **, and * denote statistical significance at the 1%, 5%, and 10% levels, respectively. *CAAR*: cumulative average abnormal return, KS test: Kolmogorov-Smirnov test - p-value ≤ 0.05 indicates normal distribution at the 5% significance level, Wx-test: Wilcoxon signed rank test as a non-parametric median test, Z-stat.: Boehmer test (parametric), t-test: skewness adjusted and bootstrapped t-test.

Table 2.3 gives our findings for the short-term risk event study. The mean and median abnormal beta shifts show a positive direction across all event windows, but we do not find any statistical significance. In contrast to the return side, the abnormal beta changes recorded are non-significant on average, but always follow a normal distribution.

Table 2.3: Risk Event Study: All Construction Industry M&A

CAAB [-x; +y]	Mean	Median	KS-test	p	Wx-test	p	t-test	p	N
0 0	0.01	0.00	0.23	0.00	-0.22	0.83	0.79	0.44	119
1 1	0.04	0.01	0.24	0.00	-0.27	0.79	0.90	0.38	119
2 2	0.07	0.01	0.26	0.00	-0.39	0.70	1.00	0.33	119
4 4	0.15	0.02	0.27	0.00	-0.48	0.63	1.03	0.32	119
7 7	0.26	0.05	0.29	0.00	-0.46	0.65	0.99	0.34	119
10 10	0.40	0.06	0.29	0.00	-0.50	0.62	1.07	0.30	119
0 1	0.02	0.01	0.26	0.00	-0.28	0.78	0.85	0.41	119
0 2	0.05	0.01	0.30	0.00	-0.30	0.76	0.95	0.36	119
0 4	0.10	0.03	0.30	0.00	-0.41	0.68	1.03	0.32	119
0 7	0.19	0.05	0.30	0.00	-0.40	0.69	1.03	0.32	119
0 10	0.28	0.05	0.29	0.00	-0.38	0.70	1.07	0.30	119

***, **, and * denote statistical significance at the 1%, 5%, and 10% levels, respectively. *CAAB*: cumulative average abnormal beta shifts, KS test: Kolmogorov-Smirnov test - p-value ≤ 0.05 indicates normal distribution at the 5% significance level, Wx-test: Wilcoxon signed rank test as a non-parametric median test, t-test: standard parametric mean test.

When we examine only the short-term perspective on M&A in the construction industry, we observe a positive shift in mean returns as the first moment of returns, and a meaningful but not significant shift in beta risk as part of the second moment. This confirms previous findings by Rottke, Schiereck, and Pauser

(2011) and Choi and Russel (2004). We find they neither over- nor underestimated the shareholder wealth effects of acquirers for a total sample analysis assuming a constant risk factor. To summarize, our expectations for first and second moment return behavior have been met. Not surprisingly, this holds true for both the return and risk sides. We find that the beta-decreasing diversification effect has likely been offset by the increased uncertainty over successfully managing a diversified entity as well as post-merger integration concerns.

Our expectations for total sample return and risk behavior are subsequently dependent on the strategic driver *transaction direction*. As noted by other scholars of financial economics, transaction direction appears to be one of the most critical success factors of construction industry M&As. Hence, we again aim to validate previous evidence on return shifts by using our sample to potentially add new insight to findings on contingent beta shifts.

Standard event study results show positive abnormal returns for horizontal, vertical, and lateral M&A. However, Table 2.4 shows that lateral M&A are especially likely to result in negative *CAARs* for the symmetric seven- and ten-day windows around the event, although these results lack statistical significance.

Table 2.4: Standard Event Study: Construction M&A along Transaction Directions

CAAR	[-x; +y]	Mean	Median		KS-test	p	Wx-test	p	Z-stat.	p	t-test	p	n
Horizontal	0 0	0.60%	0.04%		0.11	0.37	-0.68	0.50	0.97	0.33	1.17	0.10	61
	1 1	1.34%	0.67%	*	0.10	0.59	-1.71	0.09	1.91	0.06	1.89	0.01	61
	2 2	1.79%	0.09%	*	0.13	0.24	-1.66	0.10	2.17	0.03	2.17	0.01	61
	7 7	2.47%	2.50%	**	0.16	0.06	-2.10	0.04	2.15	0.03	2.22	0.01	61
	10 10	2.64%	1.12%		0.14	0.18	-1.45	0.15	2.17	0.03	1.97	0.01	61
Vertical	0 0	1.71%	1.30%	***	0.16	0.35	-2.71	0.01	2.44	0.01	2.50	0.01	32
	1 1	1.95%	0.94%	**	0.18	0.24	-2.13	0.03	2.11	0.04	2.26	0.01	32
	2 2	1.05%	0.05%		0.15	0.45	-0.37	0.71	0.75	0.46	0.87	0.50	32
	7 7	1.61%	1.63%		0.13	0.57	-0.69	0.49	1.08	0.28	1.00	0.50	32
	10 10	-0.36%	-0.25%		0.11	0.83	-0.15	0.88	-0.06	0.95	-0.19	0.50	32
Lateral	0 0	1.06%	1.31%		0.27	0.04	-1.94	0.05	0.76	0.45	0.74	0.50	26
	1 1	2.11%	1.08%	**	0.24	0.08	-1.97	0.05	1.28	0.20	1.21	0.50	26
	2 2	1.84%	0.84%		0.21	0.16	-1.00	0.32	0.93	0.35	0.95	0.50	26
	7 7	-1.73%	-1.71%		0.12	0.79	-0.39	0.69	-0.80	0.42	-0.75	0.50	26
	10 10	-2.24%	-0.66%		0.11	0.87	-0.88	0.38	-0.86	0.39	-0.92	0.50	26

***, **, and * denote statistical significance at the 1%, 5%, and 10% levels, respectively. *CAAR*: cumulative average abnormal return, KS test: Kolmogorov-Smirnov test - p-value ≤ 0.05 indicates normal distribution at the 5% significance level, Wx-test: Wilcoxon signed rank test as a non-parametric median test, Z-stat.: Boehmer test (parametric), t-test: skewness adjusted and bootstrapped t-test.

Our findings are generally in line with previous evidence, despite the fact that return shifts along transaction directions were somewhat ambiguous. Positive *CAARs* for horizontal M&A are also reported by Choi and Russel (2004), and partially by Rottke, Schiereck, and Pauser (2011). Both sets of authors strongly support our findings of positive abnormal acquirer returns for vertical M&A. However, the findings of positive return shifts for lateral M&A are in contrast to Rottke, Schiereck, and Pauser's (2011) results of significantly negative *CAARs*, and to Choi and Russel's (2004) evidence of insignificantly positive and negative *CAARs*. To summarize, we believe our findings ultimately confirm previous evidence and confirm our hypotheses with a cross-sectionally positive bias.

Our adapted risk event study reveals a new insight: that lateral M&A exhibit positive abnormal beta shifts in the symmetric seven- and ten-day windows around the event. This highlights a potential limitation of previous evidence, which neglected short-term risk dynamics. Table 2.5 provides our detailed risk event study results along transaction directions.

Table 2.5: Risk Event Study: Construction M&A along Transaction Directions

CAAB	[-x; +y]	Mean	Median	KS-test	p	Wx-test	p	t-test	p	n
Horizontal	0 0	0.01	0.00	0.30	0.00	-0.40	0.69	0.74	0.46	61
	1 1	0.05	0.01	0.32	0.00	-0.21	0.83	0.82	0.42	61
	2 2	0.11	0.01	0.34	0.00	-0.12	0.91	0.90	0.37	61
	7 7	0.41	0.04	0.39	0.00	-0.18	0.85	0.86	0.39	61
	10 10	0.59	0.01	0.40	0.00	-0.31	0.76	0.85	0.40	61
Vertical	0 0	-0.01	-0.01	0.24	0.05	-0.60	0.55	-0.24	0.82	32
	1 1	-0.02	-0.03	0.24	0.04	-0.75	0.45	-0.34	0.73	32
	2 2	-0.04	-0.05	0.26	0.02	-0.73	0.47	-0.36	0.72	32
	7 7	-0.12	-0.17	0.24	0.04	-0.93	0.35	-0.33	0.75	32
	10 10	-0.08	-0.14	0.23	0.06	-0.69	0.49	-0.17	0.87	32
Lateral	0 0	0.02	0.02	0.16	0.50	-1.51	0.13	0.83	0.42	26
	1 1	0.07	0.05	0.14	0.66	-1.46	0.14	1.13	0.27	26
	2 2	0.12	0.08	0.14	0.61	-1.59	0.11	1.21	0.24	26
	7 7	0.37	0.25 *	0.15	0.51	-1.89	0.06	1.36	0.19	26
	10 10	0.55	0.33 *	0.15	0.56	-1.89	0.06	1.49	0.15	26

***, **, and * denote statistical significance at the 1%, 5%, and 10% levels, respectively. *CAAB*: cumulative average abnormal beta shifts, KS test: Kolmogorov-Smirnov test - p-value ≤ 0.05 indicates normal distribution at the 5% significance level, Wx-test: Wilcoxon signed rank test as a non-parametric median test, t-test: standard parametric mean test.

We can confirm our hypothesis about beta stability around horizontal M&A, but we find abnormal negative beta risk shifts for vertical M&A only indicatively. Our expectations for abnormal beta increases for lateral M&A are confirmed, but only at a low level of significance.

To conclude our discussion of results across transaction direction, we have found new evidence of risk dynamics for lateral M&A, but previous findings were generally confirmed. Vertical and horizontal M&A create value for acquirer shareholders that is driven by a positive shift in the mean as the first moment of returns. For lateral M&A, any potential value creation of a positive return shift seems to be outweighed by a concurrent abnormal increase in beta.

Our next step is to investigate the effect of *transaction volume* on return and risk dynamics around construction industry M&A. We divide our total sample into three roughly equally sized subsamples: 1) thirty-nine transactions smaller than USD 40 million, 2) forty transactions between USD 40 million and USD 130 million, and 3) forty transactions larger than USD 130 million.

Interestingly, the absolute size measures of the smaller transactions are found to generate more value for acquirer shareholders than the medium-sized or larger ones. Three of the five asymmetric event windows show a high significance for positive return shifts within the small volume transactions. In theory and in actuality, non-rejectable medium-sized transactions result in zero abnormal returns. This is basically in line with findings for larger transactions that show few positive acquirer abnormal returns. Table A2.3 in the appendix shows the results for the detailed event study subsample.

Note that risk dynamics nearly mirror return dynamics in that higher returns stem from higher levels of risk, and vice versa. However, as Table A2.4 on the risk event study results shows, no statistically significant abnormal beta shifts are found. We can use the Wilcoxon signed rank test to identify statistically significant positive risk shifts for low-value transactions. However, the assumption that a normal distribution of *CABs* across event windows is justifiable, and that more relevant t-statistics will reject such significance, makes the Wilcoxon statistics somewhat questionable.

The second size factor, *relative size*, is measured as the percentage of transaction volume over acquirer market value, and returns no contradictory results. Return-induced value generation can be achieved by using relatively medium-sized M&A, where the target (transaction volume) ranges from 10%-25% of acquirer market value. Smaller and larger-sized M&A generate little return-driven value for acquirer shareholders. We do not observe any statistically significant abnormal beta changes with respect to relative size.

To summarize, this part of our discussion centered on how size measures can impact acquirer shareholder wealth effects via shifts in builders' risk-return profiles. We note that the prospect of acquisitions of less than USD 40 million that represent from 10% to 25% of an acquirer's market value is potentially best interpreted by capital market participants. This would be mainly a mean return-driven value creation, because beta remains nearly unchanged during the respective short-term event windows. We interpret these results to mean that size measures support diversification effects, in contrast to post-merger success uncertainty, but within a lower-volume range. For larger M&A, long-term success uncertainty is again likely to offset any immediate diversification benefits.

As a robustness check on our results, we next perform two analyses on potential biases. First, we focus on distinct time periods that represent consecutive economic cycles. Second, we analyze how geographic location affects M&A transactions.

We examine the risk-return profile behavior of builders around M&A transactions, and we control for three consecutive time periods, 1988-1995, 1996-2001, and 2002-2007. These time periods represent consecutive economic cycles going backward and each period comprises a sufficient amount of transactions to perform separate analyses. We find only significantly positive abnormal returns for the most recent period. We also find unusual evidence of positive *CAARs* in all three regions, Asia, Europe, and North America, but again no significant risk changes. The resulting tables are available from the authors upon request.

We then evaluate the results of our robustness analyses, and find that the most recent economic cycle (sixty-nine transactions from 2002-2007, which is more than half of our total sample) probably biased the results for our total sample standard event study positively. This could explain why our sample generally shows more continuously positive abnormal returns across transaction directions than previous studies.

2.4.3. Long-term Risk Behavior

The second part of our risk analysis examines a long-term risk perspective on (levered) equity beta and (unlevered) asset beta, as well as their dynamics from the year prior to the second year post-M&A. We pursue a twofold approach to enhance the explanatory power of the results: 1) a CAPM-based beta, obtained using standard OLS regressions, combined with 2) a GARCH (1,1)-founded beta. The beta estimates for the year prior to the M&A are determined by using the beta value thirty trading days prior to the actual transaction announcement. This may eliminate any biases through the actual event date or through advance information that may have become public beforehand.

Beta estimates for the first and second years following the M&A announcement are again determined by using the beta values from the thirty trading days after the M&A, plus 251 trading days for the first year, and 502 days for the second year. We can thus exclude the extended event window of +/- 30 trading days around the event date, and minimize any potential biasing effects from the actual event days on the long-term systematic risk estimates.

Beginning with our total sample analysis, we find no empirical evidence for equity or asset beta changes from the year prior to the M&A announcement through the second year post-M&A. However, our long-term risk analysis reveals that vertically directed M&A incur a significant decrease in acquirer asset beta up to the second year after the announcement. This finding is limited by the fact that the results apply only to the OLS beta estimates, and cannot be confirmed by the GARCH beta estimates. Nevertheless, this finding implies that business risk behavior can be distinguished from the equity risk behavior for builders pursuing vertical M&A. Adding this negative asset beta shift to the exceptionally strong positive abnormal returns (compared to horizontal and lateral M&A) highlights the value creation potentially achievable through vertical M&A. Table 2.6 shows the long-term asset beta changes.

Table 2.6: Long-term Risk Analysis: Asset Beta Dynamics along Transaction Directions

Asset Beta change in years relative to event												
Horizontal	[-x; +y]		Mean	Median		KS-test	p	Wx-test	p	t-test	p	n
	-1	1	-0.04	-0.05		0.17	0.08	-1.04	0.30	-0.83	0.41	56
OLS	-1	2	-0.05	-0.05		0.08	0.83	-0.97	0.33	-0.98	0.33	56
	1	2	-0.01	0.02		0.12	0.37	-0.05	0.96	-0.35	0.73	56
	-1	1	-0.06	-0.05		0.09	0.71	-1.26	0.21	-1.07	0.29	56
GARCH	-1	2	-0.08	0.00		0.10	0.56	-0.70	0.48	-1.19	0.24	56
	1	2	-0.01	0.02		0.12	0.34	-0.24	0.81	-0.30	0.76	56
Vertical												
	-1	1	-0.02	-0.04		0.09	0.98	-0.70	0.49	-0.55	0.58	27
OLS	-1	2	-0.12	-0.03	*	0.17	0.40	-1.66	0.10	-2.12	0.04	27
	1	2	-0.10	-0.09	**	0.10	0.91	-2.04	0.04	-2.25	0.03	27
	-1	1	-0.07	0.04		0.34	0.00	-0.84	0.40	-0.54	0.59	27
GARCH	-1	2	-0.06	0.05		0.31	0.01	-1.01	0.31	-0.40	0.69	27
	1	2	0.01	-0.02		0.13	0.71	-0.17	0.87	0.08	0.94	27
Lateral												
	-1	1	0.07	0.00		0.12	0.90	-0.68	0.50	0.90	0.38	21
OLS	-1	2	0.07	0.07		0.14	0.78	-0.82	0.41	0.88	0.39	21
	1	2	0.00	-0.02		0.23	0.19	-0.36	0.72	0.03	0.98	21
	-1	1	0.00	-0.03		0.08	1.00	-0.05	0.96	-0.05	0.96	21
GARCH	-1	2	0.02	0.05		0.14	0.75	-0.47	0.64	0.21	0.84	21
	1	2	0.03	0.04		0.09	0.99	-0.43	0.66	0.39	0.70	21

***, **, and * indicate statistical significance at the 1%, 5%, and 10% levels, respectively. KS-test: Kolmogorov-Smirnov test - p-value ≤ 0.05 indicates normal distribution at the 5% significance level, Wx-test: Wilcoxon signed rank test as a non-parametric median test, t-test: parametric mean test.

Just as in the short-term perspective, we find that *transaction volume* may again be a second relevant determinant of potential shifts in builders' risk-return profiles over the long-term. Here, we find evidence of an increase in equity beta following smaller M&A (< USD 40 million), and an equity beta decrease following medium-sized M&A (< USD 30 million). As Table A2.5 in the appendix shows, these findings are sporadic, come solely from OLS beta shifts, and are of low significance. GARCH beta changes for small transactions just miss being significant to match the OLS beta increase. The respective asset beta behavior shows no significant risk changes.

The results for *relative size* as a potential second determinant for long-term equity and asset beta behavior have little further explanatory power than those for absolute transaction volume. There is a positive equity beta shift for relatively smaller-sized M&A from the year prior to the year after the transaction at low significance. This is paralleled by a negative asset beta shift from the first to the second year post-M&A at a 5% level of significance. Again, this may provide some (anecdotal) evidence that M&A tends to drive equity and asset beta apart, but relative size does not seem to significantly impact long-term risk behavior. The tabular results are again available from the authors upon request.

Despite the fact that we obtain only meager results from our long-term equity and asset beta analyses, we perform the standard robustness analysis for consistency. Our evidence on asset beta decreases subsequent to vertical M&A may have been biased by a time or geographic factor that we would need to discredit. Table A2.6 shows asset beta decreases from 1996-2001 measured by both the OLS and GARCH models. It also shows an increase in equity beta for the 2001-2007 period at a 5% significance level by OLS, and an indicative but not statistically solid basis by GARCH. Because we believe these time period biases will cancel each other out, we infer no major effect on the asset beta decrease for vertical M&A. Our second robustness check involves a geographic focus. For Europe, there are only two significant asset beta decreases at a 10% significance level. Again, we do not note any major bias on the aforementioned asset beta decrease from this geographic focus. The table is available from the authors upon request.

2.4.4. Multivariate Regressions on Risk Behavior

For a thorough examination of integrated systematic risk behavior, we next perform several multivariate regressions. Our objective is to find supporting evidence for our total sample results and the results of our subsample univariate risk analyses. We believe the value-adding feature of this work stems primarily from the focus on the risk perspective. We thus limit our multivariate analyses to the second moment of returns.

For the short-term risk perspective, we choose the *CAAB* of the most extended event window [-10;+10] as the dependent variable. For the long-term asset and equity risk perspective, we mirror this procedure by choosing the beta changes from the year prior to the M&A announcement to the two years post-M&A. Because regressors can act selectively, we use the following variables: δ_1: lateral transaction direction; δ_2: vertical transaction direction; δ_3: transaction volume; δ_4: relative size (transaction volume divided by acquirer market value); δ_5: the 1996-2001 economic cycle; δ_6: the 2002-2007 economic cycle; δ_7: the European geographic focus; and δ_8: the North American geographic focus. Although we have analyzed the risk and return effect of all three transaction directions thus far, we do not analyze the risk effects through horizontal M&A any further in our multivariate regressions. Moreover, our previous analysis only found evidence of vertical and lateral transaction directions being (probable) risk determinants, so we concentrate primarily on these.

Our short-term-focused multivariate regression does not provide any supporting evidence for our previous results. We find significance only once for a positive δ_3 (transaction volume) at the 10% level. In addition, δ_8 (North America) is significantly negative twice at up to a 5% level. See Table A2.7 in the appendix for details.

For unlevered beta (asset beta), we find that OLS- and GARCH-based multivariate regressions on business risk determinants generally corroborate each other. However, the OLS-based regression results are somewhat better, which confirms previous evidence and adds further insight. As Table 2.7 shows, δ_1 is found to be significant twice. Models (2) and (11) suggest that lateral M&A can lead to an increase in systematic asset risk from one year prior to two years after the respective transaction. For vertical transaction direction, δ_2, models (1) and (10) have a negative coefficient, which supports the evidence from the standard long-term asset risk analysis. However, the coefficients are not statistically significant. For size measures, model (4) finds statistical significance for a positive impact of transaction volume; models (1), (5), (10), and (11) underline the strong positive impact of relative size on acquirer asset beta. Finally, models (1), (7), (10), and (11) report highly significant δ_6 coefficients, implying that the 2002-2007 time period also biases asset risk positively. The results for GARCH asset beta determinants are comparable, but they are less significant and thus not included here.

Table 2.7: Multivariate Long-term Risk Analysis: OLS Asset Beta Changes

Dependent variable OLS asset beta change [-1;+2]										
Model		1	2	4	5	7	8	9	10	11
	α	-0.15	-0.07 *	-0.05	-0.09 **	-0.13 ***	0.00	-0.03	-0.17 ***	-0.21 ***
		-1.1	*-1.9*	*-1.4*	*-2.1*	*-2.5*	*-0.1*	*-0.8*	*-2.8*	*-4.0*
Independent variables										
Lateral M&A	δ_1	0.07	0.15 *							0.13 *
		0.7	*1.6*							*1.6*
Vertical M&A	δ_2	-0.09							-0.10	
		-1.0							*-1.4*	
Transaction volume	δ_3	0.00		0.00 *						
		-0.2		*1.8*						
Relative size	δ_4	0.14 *			0.16 *				0.17 *	0.17 **
		1.8			*1.8*				*1.8*	*2.0*
1996-2001	δ_5	0.14								
		1.3								
2002-2007	δ_6	0.25 **				0.15 **			0.17 ***	0.16 **
		2.1				*2.2*			*2.6*	*2.3*
Europe	δ_7	-0.12					-0.07			
		-1.5					*-0.9*			
North America	δ_8	-0.15						-0.05		
		-1.5						*-0.5*		
Nobs.		104								
R^2 (adj.)		0.08	0.02	-0.01	0.04	0.03	0.00	-0.01	0.09	0.09
F-stat. (p-val.)		0.04	0.26	0.75	0.07	0.12	0.65	0.74	0.01	0.01
DW statistic		1.57	1.41	1.46	1.51	1.52	1.50	1.45	1.56	1.55

***, **, and * denote statistical significance at the 1%, 5%, and 10% levels, respectively; t-statistics are in italics; models (3) and (6) are excluded because the coefficients δ_2 and δ_5 are not individually statistically significant.

We perform the long-term multivariate analysis on equity beta adjustments for the OLS and GARCH betas. The OLS beta-based analysis returns a significantly positive δ_3 (transaction volume), twice returns a significantly positive δ_6 (the 2002-2007 economic cycle), and also returns a negative δ_7 (Europe). The GARCH-based equity beta multivariate regression shows a highly significantly negative δ_4 in models (1) and (5) for the relative size determinant. This indicates that increasingly large transactions lead to a decrease in equity beta up to two years after a M&A. We have not found such a relationship before. Nevertheless, we believe a size-induced magnifying diversification effect could be the explanation for this observation. Table A2.8 shows the respective figures; the OLS equity beta results table is not included here.

2.5. Conclusion

Our purpose in this paper is to reexamine the issue of acquirer wealth effects through diversification using the special case of construction industry M&A. Previous evidence by Rottke, Schiereck, and Pauser (2011) and Choi and Russel (2004) on M&A in construction has found positive acquirer wealth effects. However, these insights into wealth creation are limited to the return side, as they focus only on standard event study methodology that analyzes short-term abnormal returns. This procedure is implicitly based on the assumption of a constant beta factor.

We are eager to determine whether risk dynamics can really be neglected when deriving conclusive statements about acquirer wealth creation around M&A. As other studies on systematic risk have shown, this measure as a component of second moment returns exhibits significant differences across diversification strategies. Therefore, we concentrate on analyzing beta risk around M&A in a specific sector, the construction industry, in order to investigate the impact of potential risk changes on acquirer value creation.

In our first step, we perform a standard event study to obtain abnormal returns comparable to previous findings. Second, we analyze beta risk dynamics using an adapted risk event study for a comparable short-term method. Third, we measure long-term asset beta versus equity beta behavior to obtain a detailed view of business risk separately from financing risk. And, fourth, we performed several multivariate regressions, with the objective of validating the impact of transaction direction and selective determinants on short- and long-term risk adjustments.

Our reexamination of abnormal returns around construction industry M&A and across transaction directions found results comparable to previous studies: It confirms a decreasing contribution of vertical, horizontal, and lateral transaction directions to acquirer wealth creation. However, our sample is positively biased from strong positive abnormal returns during the 2002-2007 economic cycle. The beta risk dynamic results for the short-term window are comparable to the standard event study results. Also the results for the equity beta versus asset beta shifts in the long-term risk perspective should not be ignored. Although risk dynamic findings are selective, they are insightful as well, and add to previous evidence.

Vertical M&A have also been shown to be strongly value-increasing to acquirer shareholders based on pure mean return dynamics. Our beta risk analysis finds further evidence of a value increase through the asset beta decrease from the year prior to the M&A to two years afterward. Thus, previous evidence has underestimated the value creation to acquirer shareholders through vertical construction industry M&A.

In contrast, lateral M&A can lead to a positive abnormal equity beta risk shift in a short-term perspective, and to an asset beta increase in the long-term. Our risk event study returns a significantly increasing beta within just the twenty trading days surrounding the lateral M&A announcement. And the multivariate asset beta analysis provides further evidence that asset beta increases significantly from the year prior to two years post-M&A. These positive beta risk shifts are value-diminishing, and most probably compensate for any potentially positive abnormal returns around lateral M&A.

We find that smaller- and medium-sized transactions incur positive abnormal returns. However, the results of our analyses on *size measure* as a risk determinant are ambiguous. Our risk event study does not return any significant short-term abnormal risk changes with respect to size measures. Our long-term risk analysis finds some evidence of an equity beta increase for smaller M&A. For the small size subsample, we observe an increase in equity beta and a decrease in asset beta.

We contrast these findings with those from the multivariate regressions. The multivariate analyses on short- and long-term (OLS) risk dynamics returned significantly positive coefficients for transaction volume. Over the long run, relative size appears to influence equity beta negatively and asset beta positively. However, taken together, these results do not suggest any comprehensive effect of size measures on firm risk.

One problem is the 2002-2007 economic cycle, which affected mean returns positively and biased beta risk measures similarly. A potential bias through geographic focus appears unsystematic. Overall, however, we believe that neither time nor geographic aspects are major sources of bias for our risk-related findings.

Our ultimate conclusion is threefold: First, we confirm previous evidence on construction industry M&A and on acquirer wealth effects (concentrating on the mean as the first moment of returns) that vertical M&A is preferable to horizontal M&A, which is further preferable to lateral M&A. Second, we find that the analysis of risk behavior (which in our case is systematic risk as part of the second moment of stock returns) is value-adding to the discussion of wealth effects around M&A. Several distinct findings show that risk behavior has a considerable wealth impact – at least around construction industry M&A. Third, our findings on beta risk behavior show that the positive acquirer wealth effects around vertical M&A announcements have been underestimated thus far. We identify an additional long-term asset beta decrease, which adds to return-induced value creation. We also find short- and long-term increases in equity and asset beta for builders who pursue lateral M&A further. This underlines how this growth strategy has become critical to capital market participants.

We have two primary recommendations for future research, which reflect the limitations of our study. First, we would suggest taking a broader look at the diversity of industries to discover whether transaction directions as diversification strategies are true core value drivers. This could also help to determine whether including the beta risk perspective generally adds value to the discussion of M&A wealth creation. Second, we suggest researchers perform further testing of the applied beta risk measurement methodologies to advance their performance.

3. How Consolidation Changes the Risk Profile of a Whole Sector: Evidence from the Brewing Industry[9]

3.1. Introduction

While the wealth effects of mergers and acquisitions (M&A) have been extensively discussed in empirical M&A research, comparably few studies have given attention to the implications of M&A for company risk. This seems particularly surprising given the fact that most analyses on wealth effects utilize the market model and estimate its parameters (i.e., alpha and beta) based on pre-event-estimation period returns, assuming the parameters to be stable during the specified event window. This assumption may not necessarily hold for corporate events such as M&A, which might impact the risk characteristics of the involved firms. Conn (1985) argues that mergers in contrast to other corporate events such as earnings announcements are likely to impact a firm's operating and/or financial risk profile and thus also affect systematic risk. Consequently, the measurement of wealth effects without considering potential changes in systematic risk may lead to a biased interpretation of results. While a number of empirical studies have analyzed long-term changes in systematic risk as a result of M&A, the results are ambiguous with studies reporting positive beta changes (Langetieg, Haugen, and Wichern, 1980), negative beta adjustments (Mandelker, 1974; Lubatkin and O'Neill, 1987; Davidson, Garrison, and Henderson, 1987; Chatterjee and Lubatkin, 1990) as well as beta changes that are not significantly different from zero (Pettway and Yamada, 1986; Elgers and Clark, 1980). This mixed evidence might be based on deviating risk changes in various industries which are not detected in cross-sectional analysis. Therefore, we decide to focus on one industry with an ongoing consolidation and a global competitive environment.

Towards the turn of the last century, many industries including the brewing industry have experienced a sharp increase in M&A activity. Consolidation has been and continues to be a major trend in the sector as multi-national breweries, in an attempt to offset the decline in beer volumes in mature markets (in particular Western Europe), seek to expand their activities into new emerging markets.

---

[9] I thank Ramit Mehta for his major contribution to this study concerning the industry specific literature review and the study structure. I am also grateful to Dirk Schiereck for valuable comments and insights.

In contrast to many other sectors, the production, distribution, and marketing of beer is characterized by a relatively high fixed cost base, resulting in high levels of operational leverage (Earlam et al., 2010) providing larger brewers with material size advantages. Moreover, increased size has enabled brewers to build up a significant amount of market power (Slade, 2004), as larger brewers are able to negotiate favorable terms with their suppliers and benefit from greater bargaining power for negotiations with retail customers. Hence, it is not surprising that the global beer market today is dominated by large national/multinational companies rather than regional brewers. The four largest brewers Anheuser-Busch Inbev, Heineken, SABMiller and Carlsberg, the "big four", control about 50% of the global beer market. Despite rising market concentration, competition among the large brewing groups has remained fierce (Iwasaki, Seldon, and Tremblay, 2008). Going forward, industry experts predict that the consolidation process will continue and that the "big four" will increase their control to 75% of the global beer market (Jones, 2010). As sector debt levels are expected to decrease further, research analysts are certain that M&A will remain a major theme in the coming years as brewers will continue their quest for suitable M&A targets (Earlam et al., 2010).

In the light of these specific industry characteristics and the recent developments in the sector, the question remains open whether the M&A strategies employed by brewing companies are also reflected by capital markets in the form of abnormal risk shifts to acquiring brewing companies and their close rivals. More specifically, from a manager's perspective the question arises whether brewing companies, in addition to realizing potential revaluation of their equity, can alter systematic risk and whether it is persistently affected as a result of M&A announcements. Thus, with our analysis we aim to complement event study methodology, focusing on short-term abnormal returns. Comparable to event studies, a major concern for short-term risk measurement lies in dealing with event-induced variance. In the case of event studies, event-induced variance is best accounted for by applying parametric test-statistics that according to Harrington and Shrider (2007) use standard errors that are robust to cross-sectional variation in true abnormal returns. Advancing previous research, we analyze volatility behavior directly, measuring beta-risk dynamics around the event date. However, we separate event-induced variance from the event's persistent effect on the asset-to-market correlation coefficient. When forecasting beta around the event date we use actual daily market and asset standard deviations but the constant correlation coefficient form 41 trading days before the event. Subtracting actual from expected betas we shed light on short-term abnormal systematic risk changes.

Even though the global beer industry has gone through significant consolidation and seen a lot of M&A in recent years, empirical evidence remains scarce. Therefore, the aim of our study is also to fill this research gap and provide additional insights in M&A effects for investors and managers of beer companies. By this we contribute to the discussion of related topics around M&A, e.g., by Banal-Estañol and Seldeslachts (2011) who investigate merger failures and indentify the interaction between the pre- and post-merger processes as critical. An earlier study by Banal-Estañol and Ottaviani (2006) yet analyzed diversification effects around horizontal mergers, comparable to our brewery industry focus, finding a direct risk-sharing effect and an indirect strategic effect.

The remainder of this paper is structured as follows: Section 3.2 provides a brief overview of the relevant literature and outlines the derived hypotheses. Section 3.3 provides details on the applied methodology as well as the sample selection procedure. The following section 3.4 presents the empirical results and elaborates on the derived hypotheses. Finally, section 3.5 summarizes the findings and concludes.

3.2. Literature Review

3.2.1. Evidence on Risk Changes and Wealth Effects around M&A

Even though most studies on wealth effects of M&A utilize the market model and estimate its parameters (alpha and beta) based on a pre-event estimation period, several authors have questioned the stability of these parameters. Their general concern is based on the perception that unlike corporate events such as earnings announcements, M&A may significantly impact the operating and/or financial risk profiles of acquirer and target companies (Kiymaz and Mukherjee, 2001; Conn, 1985). While researchers have conducted several studies investigating parameter shifts subsequent to M&A, the majority of studies focus on long-term risk changes. At the same time, the studies provide mixed results:

Joehnk and Nielsen (1974) analyze the immediate and long-term effects of M&A on beta for a sample of 44 acquiring firms. They categorize their sample into various subsamples including conglomerate and non-conglomerate transactions and consider relative differences in pre-event betas between acquirer and target companies. As one of six analyzed subsamples indicates a significant change in beta between pre and post event period, Joehnk and Nielsen conclude that overall their study shows weak evidence of the responsiveness of beta to

M&A. They attribute the change in beta to be related to changes in operating and financial characteristics of the merging firms.

Based on a sample of 128 acquiring firms, Mandelker (1974) finds that the average beta increases in the pre-event period of -100 to -20 months by approximately 10% and subsequently decreases by 5% during the -20 to zero period and another 5% during the zero to +40 month period. Based on Mandelker's findings Conn (1985) concludes that cumulative abnormal returns are highly sensitive to the period chosen to estimate the parameters of the market model and therefore questions their validity.

Lubatkin and O'Neill (1987) analyze long-term changes in acquirer risk for 297 mergers, categorizing them based on the degree of the relatedness of the mergers. While the authors report significant increases in unsystematic risk for the entire sample, they provide evidence of a significant decline in systematic and total risk for related mergers. Consequently, the authors conclude that risk reduction may be a valid rationale for M&A. Davidson, Garrison, and Henderson (1987) come to a similar conclusion and attribute the risk reduction potential to merger synergies. More recently, Kiymaaz and Mukherjee (2001) confirm these results for cross-border transactions. The authors examine risk shifts for 112 transactions by US-based acquirers and report a significant decline in systematic risk in the post announcement period.

In contrast to the above-mentioned studies, that generally suggest a downward shift in the post-event beta, Langetieg, Haugen, and Wichern (1980) report a risk shift in the opposite direction. Based on a sample of 149 acquirers, the authors unexpectedly report a merger-induced increase in levels of systematic risk, total risk and unsystematic risk for the consolidated firm.

Hackbarth and Morellec (2008) argue that the sign of the risk change is dependent on the relative riskiness of acquirer and target. Using a real option model, the authors show that if the acquirer is more (less) risky than the target prior to the takeover, firm-level betas of acquirers increase (decrease) prior to the announcement and decrease (increase) in the subsequent 6 months.

On the other hand, there are also studies that report no evidence of post-event parameter changes. Based on an analysis of 59 transactions, Haugen and Langetieg (1975) conclude that there is only weak and thus no sufficient evidence that M&A lead to parameter changes. Similarly, Dodd (1980), in a study analyzing wealth effects of M&A, does not find differences in *CARs* using various estimation periods. While the author does not provide details on the conducted sensitivity tests, Connell and Conn (1993) argue that these results stand in contrast to the findings of Mandelker (1974). After analyzing a sample of 337 acquiring firms Elgers and Clark (1980) find that the systematic risk of the buyer portfolio approximates the market beta and is essentially unchanged between the pre- and post merger period. Likewise, Pettway and Yamada (1986) in a study of Japanese M&A transactions do not find sufficient evidence to explain an impact of M&A on systematic or unsystematic risk.

Amihud, DeLong, and Saunders (2002) come to a similar conclusion for cross-border bank mergers and find that systematic and total acquirer risk neither increase nor decrease as a result of a M&A transaction.

Empirical research also investigates the impact of M&A on rival firms. However, all of the reviewed studies focus on wealth effects of rival companies and do not specifically analyze implications of M&A on rival risk. With regard to wealth effects, researchers' findings show positive as well as negative effects on rivals: On the one hand, rival companies may benefit from the transaction announcement as a result of a positive signaling effect regarding industry attractiveness and potential future takeover activity (Eckbo, 1983; Song and Walkling, 2000). Moreover, a merger in the industry decreases the number of competitors and thus increases the likelihood of collusion, which may lead to greater monopoly rents to rival firms (Eckbo, 1983; Shahrur, 2005). On the other hand, rival firms may be impacted by negative competitive effects due to more intense competition in the industry from a new, more-efficient combined firm (Eckbo, 1983). Overall, the positive effects outweigh the negative competitive effects: For example, Eckbo (1985) reports positive announcement effects to rivals for horizontal transactions. Similar results are found by Song and Walking (2000) in a study including horizontal and non-horizontal transactions. These results are confirmed in more recent studies by Clougherty and Duso (2009), Fee and Thomas (2004), and Shahrur (2005).

As mentioned above, empirical research on M&A in the brewing industry remains scarce and primarily focuses on the U.S. brewing industry. The specific topics addressed in the studies focus on technological change in the sector (Kerkvliet et al., 1998), the sector's tendencies towards concentration (Lynk, 1985), (Adams, 2006), the determinants and motives for horizontal M&A (Tremblay and Tremblay, 1988), as well as competition in the industry (Horowitz and Horowitz, 1968). More recently, Ebneth and Theuvsen (2007) analyze the short-term value effects of M&A to acquirers using event study methodology. Based on a sample of 29 cross-border transactions involving European acquirers from 2000-2005, they find insignificant positive acquirer returns. In a companion paper, Mehta and Schiereck (2011) challenge these results and report significant positive abnormal returns to acquirers. Moreover, they find a significant negative effect to close rivals missing out on a potential M&A opportunity.

3.2.2. Contribution to the Literature and Hypotheses

With this paper we aim to contribute to existing literature in two ways. First, we provide a comprehensive analysis of short-term risk effects of M&A to acquirers in the brewing industry using a global dataset that in addition to cross-border transactions also includes domestic transactions. This enables us to specifically build and investigate relevant subsamples and determine key drivers of systematic risk adjustments. Second, we specifically analyze abnormal risk shifts of M&A transactions by the "big four" as well as the competitive effects of missing out on a transaction.

As pointed out above, previous studies report mixed results. Given the particularities of the brewing industry and its development over the last few years, we expect M&A in the sector to be a feasible measure to realize synergy and efficiency gains. However, given the fairly concentrated market structure, where it is becoming increasingly hard to find suitable targets, we believe that overall horizontal diversification potential is limited. While certain transactions might have an unexpected risk-reducing effect, for the total sample we do not expect to find a significant abnormal change in systematic risk.

In order to determine potential drivers of systematic risk adjustments, a number of hypotheses concerning the impact of deal, acquirer, and target characteristics are formulated.

Geographical scope: Connell and Conn (1993) investigate abnormal returns for 73 mergers between US and UK firms using three different estimation periods (pre-event, post-event, and pooled). The authors find that abnormal returns are highly sensitive to the estimation period chosen for the market model. More specifically, they report lower abnormal returns when a post-merger estimation period is employed and attribute 85% of the difference to decline in alphas. On the other hand, Fatemi (1984), while not specifically investigating cross-border mergers, finds a negative relationship between the international involvement of a bidder and the post-merger beta. As pointed out above, Kiymaz and Mukherjee (2001) confirm these results and report a significant decline in systematic risk in the post event period for cross-border transactions by US acquirers. The authors conclude that cross-country diversification provides risk reduction and thus value-enhancing opportunities, which cannot be replicated by investors using portfolio diversification.

In case of the brewing industry, cross-border transactions have been of particular relevance as many brewers have looked to diversify into foreign markets given the decline in beer volumes in many mature markets. In particular, brewers have looked at acquisition opportunities in emerging markets such as Asia, Eastern Europe, and Latin America, which are still relatively fragmented and offer growth opportunities for beer volumes. From a risk perspective, we assume cross-border M&A to be a feasible measure for brewers to diversify their earnings streams and thus expect to find a significant risk-reducing effect of cross-border M&A on acquirers, in particular when acquired targets are based in emerging market economies. Moreover, we expect to find significant differences between cross-border and domestic transactions.

Transaction value and relative transaction size: The brewing industry has seen a significant increase in transaction values in recent years. Given the particular characteristics of the sector, increased company size provides brewers with a material competitive advantage. Consequently, it can be argued that the acquisition of big targets will significantly contribute to the success of a transaction due to greater potential for revenue and cost synergies. From a risk perspective, a larger target may provide a greater diversification opportunity. On the other hand, the integration of larger targets is likely to be more difficult than for small targets (Hawawini and Swary, 1990) and thus potentially also more risky. In addition to target size, the size of the acquirer may also influence the success of a transaction. Moeller, Schlingemann, and Stulz (2003) argue that managers of larger companies are more likely to overestimate their own abilities. Due to the fact that larger companies may benefit from bigger cash reserves and also might need to face fewer hurdles in the execution of M&A transactions, the authors argue that managers of larger companies are more likely to engage in M&A transactions that are not always beneficial to the company.

Transaction date: Over the last few years, the brewing industry has experienced a sharp increase in M&A activity and seen strong industry consolidation. As a consequence, the competitive landscape has materially changed as the "big four" today control more than 50% of global beer volumes. While the sector is expected to further consolidate in the future, it is becoming increasingly difficult to find suitable targets (Gibbs, Webb, and Dhillon, 2010). We assume the changes in market structures and concentration to have an impact on acquirer returns and expect to find significant differences in risk effects over time.

As already mentioned above, the global beer market is dominated by Anheuser-Busch Inbev, SABMiller, Heineken, and Carlsberg, the "big four". In a companion paper, analyzing M&A announcement effects, Mehta and Schiereck (2011) specifically analyze rival effects among the "big four" and find significantly positive abnormal returns if one of the "big four" announced a M&A acquisition, while the remaining three rivals, missing out on a M&A opportunity experienced negative abnormal returns. Similarly, we expect to find a stabilizing effect if one of the "big four" engages in a M&A transaction and thus predict a decline in systematic risk. Given the strong competition among the "big four" we also expect to find an impact on the risk levels of the remaining three rivals as they are put into a disadvantageous position by missing out on a diversification and growth opportunity in a fairly concentrated market.

3.3. Data and Methodology

3.3.1. Data Selection

The sample of mergers and acquisitions for the risk event study is drawn from the Securities Data Corporation (SDC)/Thomson One Banker Deals database and the Merger Market M&A database. It includes all world-wide M&A events announced between January 1st, 1991, and December 31st, 2010. The total number of M&A deals is reduced to yield only those transactions meeting the following criteria:

i. At the time of the transaction, acquirer and target companies both had active operations in the brewing industry.

ii. The acquiring company has been publicly listed for at least 250 days prior to the announcement of the transaction.

iii. The total transaction value accumulates to at least USD 50 million.

iv. The completion of the transaction leads to a change of control in the target; prior to the announcement of the transaction the bidder holds less than 50% in the target company, following the transaction the bidder obtains a controlling stake in the target company.

v. The transaction has been successfully completed.

In addition, the transactions were validated by a press research using the Factiva database as well as company websites in order to ensure that all transactions are horizontal and the announcement dates provided by the databases are correct. Moreover, acquirers with multiple transactions on the same day were removed from the dataset. The described selection criteria result in a final sample of 75 transactions. The frequency distribution of the transactions over time is provided in Table 3.1. While the number of transactions is spread fairly even from 1998 onwards, the average transaction size varies strongly from USD 50 million to USD 11,973 million due to a number of high-profile transactions such as InBev's acquisition of Anheuser-Busch (USD 52 billion), Heineken and Carlsberg's takeover of S&N (USD 15 billion) and Heineken's recent acquisition of Femsa Cerveza (USD 5.7 billion). In terms of geography more than 75% of the transactions involve acquirers that are based in Europe.

The relevant daily stock prices, market capitalizations, and local market indices for acquirers were downloaded from the Thomson Datastream database. Acquirer returns are calculated using the Datastream Total Return Index, which adjusts the closing share prices for dividend payments as well share issuances or repurchases.

Table 3.1: Sample Overview: Descriptive Statistics

			Avg. Trans. Val.	Trans. Val.	Acquirer Region - Number of Deals			
Year	Deals	(%)	(USD mil.)	(USD mil.)	Europe	Americas	Asia	RoW
2010	1	1.3	5,700	5,700	1			
2009	4	5.3	737	2,946	2		2	
2008	6	8.0	11,973	71,835	6			
2007	5	6.7	405	2,025	1	2	1	1
2006	6	8.0	332	1,991	6			
2005	8	10.7	785	6,281	8			
2004	9	12.0	875	7,873	6	2	1	
2003	6	8.0	582	3,489	6			
2002	6	8.0	1,526	9,157	5	1		
2001	4	5.3	599	2,396	3	1		
2000	6	8.0	514	3,086	5		1	
1999	4	5.3	495	1,981	3	1		
1998	4	5.3	268	1,071	4			
1997	0	0.0	0	0				
1996	1	1.3	65	65		1		
1995	1	1.3	991	991	1			
1994	1	1.3	134	134		1		
1993	1	1.3	155	155				1
1992	1	1.3	79	79	1			
1991	1	5.8	50	50	1			
Sum	75	100.0	1,617.4	121,305.0	59	9	5	2

Sample includes all M&A transactions between 1998 and 2010 as specified above. Table 3.1 shows the frequency distribution of M&A transactions between 1998 and 2010 with total and average transaction values in USD mil. Additionally, details on acquirer region are provided.

3.3.2. Methodology of the Systematic Risk Analysis

In order to determine and measure risk shifts we apply an event study methodology that has been advanced to measure short-term shifts in beta that exceed or fall below expectations. The mathematical intuition is highly comparable to the standard event study. Our risk event study is composed of three basic steps: first, measuring daily acquirer betas, second, estimating expected betas from before M&A announcements, and third, subtracting measured betas from expected betas to obtain abnormal beta shifts, *AB*. Cumulating *ABs* within event windows and determining the arithmetic mean across events finally returns the cumulative average abnormal beta shifts, *CAAB*. Intuitively obvious to scholars of financial economics measuring time-variant beta breaches the basic assumption of the CAPM stating that beta is stable over time. But, financial economists like Jagannathan and Wang (1996), Collins, Ledolter, and Rayburrn (1987), Bos and Newbold (1984), Sunder (1980), and Fabozzi and Francis (1978) have argued against the assumption of beta stability and showed that beta is actually stochastic and thus conditional.

To determine time-varying volatility and beta measures we apply the generalized autoregressive conditional heteroscedasticity (GARCH) approach. It is one of the most commonly applied techniques by researchers to observe changes in risk levels. The methodology was first introduced by Engle (1982), refined by Bollerslev (1986) and further advanced to a Multivariate GARCH (M-GARCH) by Bollerslev (1990). See Giannopoulos (1995) for an empirical implementation of GARCH models where he finds evidence that such volatility modeling techniques cannot only be used to measure total risk but also systematic and unsystematic components. Brooks, Faff, and McKenzie (1998) investigate the forecasting accuracy of three techniques regarding time-variant beta: a M-GARCH, a time-varying beta market model suggested by Schwert and Seguin (1990), and the Kalman filter for a sample of returns on Australian industry portfolios. Furthermore, Faff, Hillier, and Hillier (2000) applied an M-GARCH approach to model time-varying beta in their study discussing alternative beta risk modeling techniques. They define an additional assumption of a constant correlation coefficient between the returns of an asset and respective market returns. The behavior of this correlation coefficient is core to our risk event study determining *CAABs*.

We apply a simple GARCH (1,1) model. As shown by Equation (3.1) it determines a specific asset's time-varying volatility in terms of the variance rate, σ_{it}^2, as a weighted sum of a long-term average variance, Φ_i, the previous day squared return, $u_{i,t-1}^2$, and the previous day variance, $\sigma_{i,t-1}^2$. The weighting factors, γ_1, γ_2, and γ_3 are determined applying the iterative Maximum Likelihood technique as discussed by Hull (2006).

$$\sigma_{it}^2 = \gamma_1 \Phi_i + \gamma_2 u_{i,t-1}^2 + \gamma_3 \sigma_{i,t-1}^2 \tag{3.1}$$

Obtaining beta measures as next step, we apply the M-GARCH model that observes the actual betas as product of a constant correlation coefficient, ρ_{im}, between the specific acquirer returns and the respective market returns, and the time-varying acquirer standard deviation, σ_{it}, divided by the time-varying market standard deviation, σ_{mt}.

$$\beta_{it} = \frac{\rho_{im}\sigma_{it}}{\sigma_{mt}} \tag{3.2}$$

Finishing the above elucidated empirical groundwork, we have the following data at hand: Daily volatility measures for all brewery industry acquirers and respective market indices; daily correlation coefficients for all acquirers with their corresponding market indices based on a one year lag structure; and daily beta measures for all acquirers.

Equivalent to beta for the standard return event study we utilize the correlation coefficient 41 trading days prior to the event (i.e., 21 trading days or one month before the start of the maximum event window [-20;+20]) for our systematic risk estimates, $E(\beta_{it})$. The correlation coefficient is determined over a 250 trading day period comprising one calendar year. Acquirer and market volatility measures remain time-variant as the market return does in the standard event study. The abnormal beta shift, *AB*, is then determined as difference between expected and measured beta from up to ten days before to up to ten days after the event.

$$AB_{it} = E(\beta_{it}) - \beta_{it} = \frac{\rho_{im}\sigma_{it}}{\sigma_{mt}} - \beta_{it} \tag{3.3}$$

Cumulating *ABs* along the differently sized event windows we obtain cumulative abnormal beta shifts, *CAB*. The arithmetic mean of *CABs* across the number of events, *n*, per analysis returns the cumulative average abnormal beta shifts, *CAAB*. The Kolmogorov-Smirnov test is used to determine if the *CABs* are normally distributed and hence to decide whether to test the *CAAB* means and medians for difference to zero with parametric t-tests or non-parametric Wilcoxon signed rank tests.

Table 3.2 provides an overview on average market and brewery risk measures such as variance, correlation coefficient, and beta. Worth mentioning is the by far higher average brewery variance than market variance. In addition, more specific brewery risk adjustments with respect to the event-window effects of variance and correlation on beta can be directly observed. The correlation coefficient drops from 41 days prior the event to the event date, whereas the variance increases in parallel. In sum, beta rises and this upward trend is significant at the 10% level, even though shifts in correlation and variance are not statistically significant. The increase in brewery variance towards the event date and thereafter is what we refer to as event-induced variance. On the other hand, we define an event's effect on the correlation coefficient as a more persistent effect on beta and consequently focus our analysis thereon.

Table 3.2: Total Sample Risk Characteristics

		Correlation	Variance	Beta		N
1997-2010	market ind.	-	0.00028	-		16
average	breweries	0.423	0.00045	0.563		26
Event window	-40	0.454	0.00036	0.634		75
	Event date	0.449	0.00038	0.666		
	20	0.450	0.00041	0.665		
Adjusments *(t-stat.)*	-41/0	-0.006	0.00003	0.033	*	75
		-1.0	*0.9*	*1.6*		
	0/20	0.001	0.00003	-0.001		
		0.2	*0.6*	*-0.1*		
	-41/20	-0.006	0.00005	0.031	*	
		-0.8	*1.1*	*1.6*		

***, **, and * indicate statistical significance at the 1%, 5%, and 10% levels, respectively. t-test: std. parametric mean test. Contents of "Event-window" and "Adjustments" sections specify brewery risk measures in relation to respective market indices.

3.4. Results of Risk Analysis around M&A

3.4.1. Acquirer Short-term Systematic Risk Changes

Figure 3.1 (next page) reports the short-term abnormal risk effects of M&A transactions on the total sample of acquirers in the brewing industry. The results show that for the total sample all daily average abnormal beta changes *(AAR)* are negative. We observe a strong increase in abnormal daily beta descents beginning from the event date though decreasing towards the end of the event window. Nevertheless, as the results in Table A3.1 in the appendix show, the observable daily risk adjustments hardly lead to any significant *CAABs*. These findings are in line with our predictions. In the following, we will analyze relevant subsamples to gain further insight on a potential relationship between M&A announcements and beta changes.

Geographical scope: In order to assess the impact of geographical diversification on acquirer beta changes we analyze domestic transactions, cross-border transactions and emerging market transactions (i.e., transactions where the acquired targets are based in emerging market economies). The results are provided below, Table 3.3 and in the appendix, Table A3.2. In line with our predictions we find a significant decrease in beta for cross-border transactions across almost all examined event windows. Acquirers experience a significant abnormal beta reduction of 0.55 and 0.29 for the [-20; 20] and [-10; 10] event window, respectively. These results suggest that cross-border transactions seem to have a stabilizing effect on acquirer systematic risk. We find similar results for cross-border transactions with targets in emerging market economies (Asia ex Japan, Latin America, and Eastern Europe). *CAABs* are significant for all examined event windows. For the [-20; 20] and [-10; 10] event windows we find negative beta changes of -0.51 and -0.23, respectively, both significant at the 5% level. In contrast, we find no impact on systematic risk for domestic transactions (see Table A3.2). While these results suggest a relationship between the geographical scope of a transaction and systematic risk, we will challenge these results with a multivariate regression analysis.

Figure 3.1: Daily Acquirer AABs throughout the Event Window

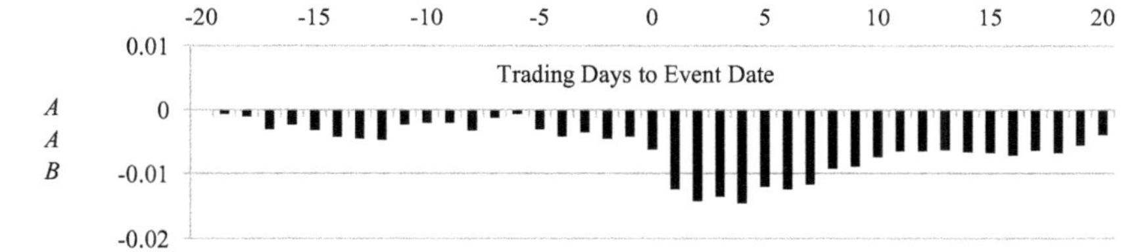

Table 3.3: Acquirer CAAB by Geographical Scope

CAAB	[-x; +y]	Mean	Median	KS-test	p	Wx-test	p	t-test	p	n
Cross-Border M&A	0 0	-0.01	-0.01	0.11	0.53	-1.50	0.13	-0.02	0.28	51
	1 1	-0.03	-0.03 *	0.11	0.56	-1.80	0.07	-0.07	0.16	51
	5 5	-0.12	-0.15 **	0.10	0.59	-1.92	0.05	-0.28	0.12	51
	10 10	-0.23	-0.29 **	0.11	0.50	-2.05	0.04	-0.53	0.13	51
	20 20	-0.42	-0.55 **	0.11	0.54	-2.09	0.04	-0.96	0.12	51
	0 1	-0.02	-0.02 *	0.13	0.36	-1.82	0.07	-0.05	0.12	51
	0 10	-0.19 *	-0.16 **	0.10	0.63	-2.02	0.04	-0.40	0.08	51
	0 20	-0.34	-0.27 **	0.09	0.78	-1.92	0.05	-0.76	0.11	51
Emerging Market M&A	0 0	-0.01	-0.01 *	0.16	0.31	-1.70	0.09	-0.03	0.17	34
	1 1	-0.04	-0.03 **	0.17	0.24	-1.97	0.05	-0.10	0.11	34
	5 5	-0.17 *	-0.08 **	0.16	0.33	-2.15	0.03	-0.36	0.08	34
	10 10	-0.32 *	-0.23 **	0.17	0.22	-2.27	0.02	-0.67	0.07	34
	20 20	-0.58 **	-0.51 **	0.14	0.51	-2.40	0.02	-1.17	0.05	34
	0 1	-0.04 *	-0.02 **	0.17	0.27	-1.94	0.05	-0.08	0.08	34
	0 10	-0.26 **	-0.16 **	0.19	0.13	-2.37	0.02	-0.50	0.04	34
	0 20	-0.48 **	-0.28 **	0.19	0.14	-2.32	0.02	-0.93	0.04	34

***, **, and * indicate statistical significance at the 1%, 5%, and 10% levels, respectively. *CAAB*: cumulative average abnormal beta shift, KS test: Kolmogorov-Smirnov test - p-value ≤ 0.05 indicates normal distribution at the 5% significance level, Wx-test: Wilcoxon signed rank test as median test, t-test: std. parametric mean test.

Transaction value and relative transaction size: Table 3.4 presents the *CAABs* for the 30 largest transactions. In line with our expectations we do not find any abnormal beta changes for the 30 smallest transactions and hence leave them out. On the other hand, we report significant negative abnormal beta changes for large transactions across all analyzed event-windows. For the [-20; 20] and [-10; 10] event windows we find negative beta changes of -0.77 and -0.42, respectively, both significant at the 5% level. The results are in line with our predictions and suggest a potentially negative relationship between transaction value and beta. Again, we will challenge the risk event study results in the multivariate analysis.

Table 3.4: Acquirer CAAB of 30 Largest M&A by Transaction Value

CAAB [-x; +y]	Mean		Median		KS-test	p	Wx-test	p	t-test	p	n
0 0	-0.01		-0.01	*	0.15	0.45	-1.70	0.09	-0.03	0.12	30
1 1	-0.06	**	-0.06	**	0.15	0.46	-2.09	0.04	-0.12	0.04	30
2 2	-0.11	**	-0.10	**	0.14	0.58	-2.13	0.03	-0.21	0.03	30
5 5	-0.24	**	-0.21	**	0.12	0.77	-2.27	0.02	-0.45	0.03	30
10 10	-0.43	**	-0.42	**	0.11	0.81	-2.27	0.02	-0.83	0.04	30
20 20	-0.73	**	-0.77	**	0.11	0.78	-2.23	0.03	-1.47	0.05	30
0 1	-0.04	**	-0.04	**	0.18	0.25	-2.07	0.04	-0.09	0.04	30
0 2	-0.08	**	-0.06	**	0.17	0.34	-2.13	0.03	-0.15	0.03	30
0 5	-0.17	**	-0.12	**	0.14	0.58	-2.25	0.02	-0.33	0.03	30
0 10	-0.32	**	-0.22	**	0.15	0.45	-2.25	0.02	-0.62	0.04	30
0 20	-0.56	*	-0.55	**	0.14	0.57	-2.07	0.04	-1.17	0.07	30

***, **, and * indicate statistical significance at the 1%, 5%, and 10% levels, respectively. *CAAB*: cumulative average abnormal beta shift, KS test: Kolmogorov-Smirnov test - p-value ≤ 0.05 indicates normal distribution at the 5% significance level, Wx-test: Wilcoxon signed rank test as non-parametric median test, t-test: std. parametric mean test. The top 30 M&As by transaction value include transactions between USD 400 mil. and 53 bil.

Table A3.3 reports the *CAABs* for the 30 smallest and 30 largest transactions by relative transaction size, which is determined as transaction value divided by acquirer market value. While we find significant beta shifts in some event windows, overall there does not seem to be enough evidence to suggest a potential relationship between relative transaction size and systematic risk. Nonetheless, the relative size of transactions will be considered as part of the multivariate regression analysis.

Transaction date: As pointed out in the literature review, the structure of the brewing industry has materially changed in the course of the last years. Table A3.4 presents changes in systematic risk for transactions from 1991-2003 and 2004-2010. We find insignificant negative beta changes for both subsamples. These results are similar to the results for the entire sample and do not suggest an impact of transaction date on systematic risk.

As already mentioned in section 3.2, the "big four" dominate the global brewing industry. Table 3.5 presents the *CAABs* for transactions by the "big four". We find significant negative abnormal beta changes across all examined event windows. For the [-10; 10] and [-20; 20] event windows we find negative *CAABs* of -0.35 and -0.62, respectively. These results are in line with our predictions. It should be noted that all "big four" transactions in the sample were cross-border transactions. Overall, it appears justified to conclude that the "big four" have been able to reduce systematic risk by engaging in M&A.

Table 3.5: Acquirer CAAB of "Big Four"

CAAB [-x; +y]	Mean		Median		KS-test	p	Wx-test	p	t-test	p	n
0 0	-0.01		-0.02	*	0.15	0.36	-1.78	0.08	-0.03	0.31	36
1 1	-0.04		-0.05	**	0.13	0.52	-2.06	0.04	-0.09	0.17	36
2 2	-0.06		-0.08	**	0.14	0.41	-2.04	0.04	-0.15	0.14	36
5 5	-0.15		-0.19	**	0.13	0.53	-2.15	0.03	-0.34	0.13	36
10 10	-0.28		-0.35	**	0.15	0.33	-2.18	0.03	-0.65	0.13	36
20 20	-0.51		-0.62	**	0.13	0.54	-2.22	0.03	-1.16	0.12	36
0 1	-0.03		-0.04	**	0.14	0.43	-2.07	0.04	-0.07	0.13	36
0 2	-0.05	*	-0.05	**	0.15	0.37	-2.17	0.03	-0.11	0.10	36
0 5	-0.12	*	-0.09	**	0.15	0.35	-2.29	0.02	-0.25	0.09	36
0 10	-0.23	*	-0.21	**	0.16	0.30	-2.23	0.03	-0.49	0.08	36
0 20	-0.42	*	-0.53	**	0.12	0.61	-2.18	0.03	-0.94	0.10	36

***, **, and * indicate statistical significance at the 1%, 5%, and 10% levels, respectively. *CAAB*: cumulative average abnormal beta shift, KS test: Kolmogorov-Smirnov test - p-value ≤ 0.05 indicates normal distribution at the 5% significance level, Wx-test: Wilcoxon signed rank test as non-parametric median test, t-test: std. parametric mean test.

3.4.2. Rival Short-term Systematic Risk Changes

Table 3.6 presents the rival effects among the "big four" i.e. the abnormal beta changes for those three companies of the "big four" missing out on a potential M&A opportunity when one of their close rivals announces a M&A transaction. In line with our predictions, we find a significant impact on systematic risk of rival companies. While average and median cumulative beta changes are positive across all analyzed event windows, we find significant positive *CAABs* for a number of the analyzed event windows with significance at the 5% and 10% level. In order to detect potential drivers of abnormal rival beta changes we specifically analyze various subsamples similar to our acquirer analysis (see Tables A3.5, A3.6).

Table 3.6: Rival CAAB among "Big Four"

CAAB [-x; +y]	Mean	Median		KS-test	p	Wx-test	p	t-test	p	N
0 0	0.00	0.01	**	0.14	0.03	-2.03	0.04	-0.01	0.49	96
1 1	0.01	0.02	**	0.15	0.02	-1.93	0.05	-0.02	0.51	96
2 2	0.02	0.03	*	0.16	0.01	-1.86	0.06	-0.04	0.56	96
5 5	0.03	0.05	*	0.16	0.01	-1.70	0.09	-0.09	0.65	96
10 10	0.03	0.09		0.17	0.01	-1.46	0.14	-0.17	0.74	96
20 20	0.07	0.15		0.16	0.02	-1.47	0.14	-0.30	0.71	96
0 1	0.01	0.01	**	0.15	0.03	-2.01	0.04	-0.01	0.46	96
0 2	0.01	0.02	*	0.14	0.04	-1.82	0.07	-0.02	0.54	96
0 5	0.01	0.04		0.14	0.03	-1.53	0.12	-0.05	0.66	96
0 10	0.02	0.05		0.13	0.07	-1.25	0.21	-0.10	0.80	96
0 20	0.03	0.04		0.13	0.08	-0.98	0.33	-0.20	0.79	96

***, **, and * indicate statistical significance at the 1%, 5%, and 10% levels, respectively. *CAAB*: cumulative average abnormal beta shift, KS test: Kolmogorov-Smirnov test - p-value ≤ 0.05 indicates normal distribution at the 5% significance level, Wx-test: Wilcoxon signed rank test as non-parametric median test, t-test: std. parametric mean test.

Geographical scope: We find significant positive abnormal beta shifts for rivals if a transaction is announced in a mature market as shown in Table 3.7. Beta changes are significant at the 5% or 10% level across almost all analyzed event windows. In contrast to most emerging market economies where beer markets are generally relatively fragmented, most mature beer markets are rather concentrated and dominated by few market players. Consequently, the consolidation potential in mature markets is generally smaller, hence providing brewers with fewer diversification opportunities. Thus, a possible explanation for the increase in systematic risk for rivals may be the fact that they missed out on a potential consolidation opportunity in a fairly concentrated market while at the same time a relevant competitor strengthened its position in the same market. We will verify these results in the upcoming multivariate regression analysis. Since we do not find any abnormal risk changes for acquisitions regarding emerging markets, we do not report respective statistics.

Transaction value and relative transaction size: While we find weak evidence for significant positive risk shifts for rivals for the smallest 30 transactions by transaction value (see Table A3.5), we do not believe that the risk event study provides sufficient evidence to explain a potential relationship between transaction value or relative transaction size and rival risk shifts.

Transaction date: We find significant abnormal positive risk shifts for rivals for transactions between 1991 and 2003 (see Table A3.6). The results are significant at the 5% or 10% level across almost all analyzed event windows. On the other hand, we do not find abnormal risk shifts for transactions between 2004 and 2010. The results are in line with our analysis on the geographical scope of transactions, as the majority of transactions between 1991 and 2003 (>54% of transactions) involved targets in mature markets, while most of the transactions between 2004 and 2010 (>75% of transactions) involved targets in comparably fragmented emerging market economies. Another possible explanation for the results could be seen in the emergence of the "big four" as well-diversified global players over time. Consequently, it could be argued that rivals were more affected by competitors' M&A activities while they were still comparably small and not as well diversified. Again the results will be verified in the multivariate regression analysis.

Table 3.7: Rival CAAB among "Big Four" in Mature Markets

CAAB [-x; +y]	Mean		Median		KS-test	p	Wx-test	p	t-test	p	n
0 0	0.02		0.02	*	0.14	0.46	-1.85	0.06	0.00	0.06	34
1 1	0.05	*	0.05	**	0.12	0.65	-1.96	0.05	0.00	0.06	34
2 2	0.08	**	0.08	*	0.12	0.66	-1.85	0.06	0.00	0.05	34
5 5	0.16	**	0.13	*	0.12	0.68	-1.87	0.06	0.00	0.05	34
10 10	0.27	*	0.22	*	0.14	0.50	-1.75	0.08	-0.02	0.07	34
20 20	0.45	*	0.43	*	0.15	0.43	-1.75	0.08	-0.03	0.07	34
0 1	0.03	**	0.04	**	0.14	0.48	-2.01	0.04	0.00	0.05	34
0 2	0.05	**	0.06	**	0.14	0.52	-2.03	0.04	0.00	0.05	34
0 5	0.09	*	0.09	*	0.12	0.69	-1.68	0.09	-0.01	0.09	34
0 10	0.13		0.14		0.12	0.62	-1.55	0.12	-0.05	0.14	34
0 20	0.22		0.22		0.09	0.89	-1.38	0.17	-0.12	0.19	34

***, **, and * indicate statistical significance at the 1%, 5%, and 10% levels, respectively. *CAAB*: cumulative average abnormal beta shift, KS test: Kolmogorov-Smirnov test - p-value ≤ 0.05 indicates normal distribution at the 5% significance level, Wx-test: Wilcoxon signed rank test as non-parametric median test, t-test: std. parametric mean test.

3.4.3. Summary of Acquirer and Rival Risk Changes

The risk event study analysis of acquirers and rivals has shown various abnormal risk changes around the announcement of M&A transactions. Thereby, two tendencies can be identified: First, acquirers could realize on average significant abnormal descents in systematic risk around the event date if certain transaction characteristics such as cross border M&A, target in emerging market, or market leading acquirer (big four) were observed. Second, rivals showed significant abnormal increases in systematic risk throughout the event window – especially, if missing out on targets in mature markets where respective M&A largely took place during the 1991 to 2003 time period. Figure 3.2 illustrates the accumulation of daily average abnormal beta changes from 20 trading days before the transaction [-20;-20] to the end of the entire event window [-20;+20]. The

CAABs at the end of the event window can also be found in the respective tables where the significance levels are indicated.

Figure 3.2: Cumulative Acquirer AABs throughout the Event Window

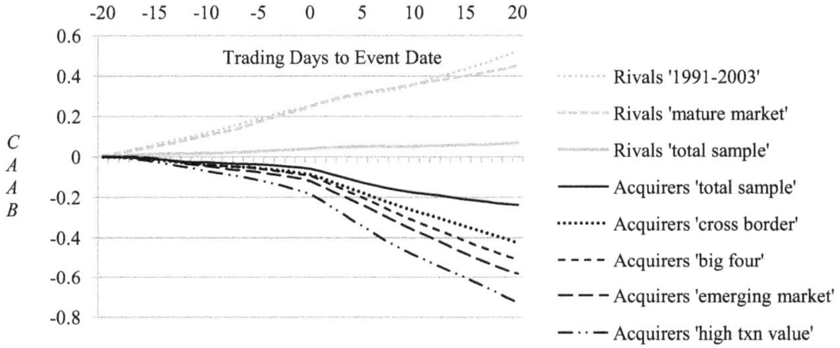

Overall, the risk event study results support our insights gained from the literature review. The abnormal beta descent for the total acquirer sample is not significant, but the risk diversifying effect of cross border M&A and by that especially the acquisition of targets in growing emerging markets confirms our expectations. Furthermore, large transactions emerge as a major driver for decreasing beta risk, supporting the argument for extended cost synergy potential in the trade-off to an increased risk of a successful integration of a larger target company. The risk-reducing effect of "big four" transactions, which coincide significantly with cross border, emerging market, and large transaction value M&A, serves as evidence for their shareholder value enhancing growth strategy.

Missing out on transactions results in a significant abnormal beta risk increase for rival breweries which is further enhanced if M&A take place in mature markets. These findings also confirm our expectations of rivals being put into a disadvantageous position having reduced their diversification potential in a fairly contracted market.

3.4.4. Determinants of Abnormal Risk Changes

In order to obtain further insights on potential drivers of systematic risk changes as yet indicated by the risk event study, two cross-sectional regressions are performed on the cumulative abnormal beta changes to acquirers and rivals.

Our first multivariate regression analysis includes up to six variables to represent the parameters, which have already been analyzed individually in the risk event study subsamples. Equation 3.4 shows the respective model with the acquirer *CAB* of the [0;+5] event window as dependent variable. The independent variables are discussed in the following paragraphs.

$$CAB_{acqu.} = \alpha + \delta_1 * CB + \delta_2 * EM + \delta_3 * B4 + \delta_4 * TV + \delta_5 * RS + \delta_6 * Yr \qquad (3.4)$$

Geographical scope: The results of the risk event study suggested that the announcement of cross-border transactions, when compared to domestic transactions, might lead to a significant reduction in systematic risk. At the same time, the results suggested a significant decline in systematic risk if targets were based in emerging market economies (Latin America, Eastern Europe, and Asia excluding Japan). Both effects are included in the regression model using the dummy variables *CB* for "cross border M&A" and *EM* for "target emerging market".

"Big four" transaction: Risk event study subsample results also showed significant negative cumulative abnormal beta shifts for transactions announced by the "big four". We test these results by including a dummy variable *B4* for transactions by the "big four".

Transaction value and relative transaction size: In line with our predictions, the risk event study suggests a potential negative relationship between transaction value and beta. On the other hand, we did not find enough evidence suggesting an impact of relative transaction size (transaction value / acquirer market value) on systematic risk. In order to verify these results, transaction value, *TV*, and relative transaction size, *RS*, are included as variables in the regression model.

Transaction date: As pointed out above, the structures of the global beer market have materially changed in recent years. In the risk event study subsample analysis we did not find significant beta changes for any time period analyzed. Nonetheless, we include the year, *Yr*, of the M&A announcement in the regression model to again scrutinize this potential driver.

Table 3.8 presents the results of four regression models (M1-M4) on the *CABs* for the [0;5] event window. This specific event window has been chosen because it is the only one that showed a significant abnormal average beta change for acquirers within the risk event study. If we may expect any significant result in the respective multivariate analysis, it will most definitely be observable within the [0;5] event window. The regression models differ with regard to the independent variables included. Overall, none of the models is statistically significant (based on F-statistics). The explanatory power is low with an adjusted R-squared ranging between -3% and 2%. Autocorrelation issues can be ruled out due to Durbin-Watson-statistics near two in both cases. The constant, alpha, is insignificant for all four models. Overall, these findings are in line with our expectations and confirm the largely insignificant beta changes determined for the total sample in the risk event study.

Table 3.8: Multivariate Regressions on Acquirer CABs

Dependent variable	*CAB* [0;5]				
Model		M1	M2	M3	M4
Intercept	α	2.49	0.02	-0.02	-0.05
		0.1	*0.4*	*-0.4*	*-1.1*
Independent variables					
Cross border M&A	δ_1	-0.03			
		-0.3			
Target emerging market	δ_2	-0.13	-0.14 *	-0.11	
		-1.5	*-1.6*	*-1.2*	
Acquirer size (Big Four)	δ_3	-0.02			
		-0.2			
Transaction size	δ_4	0.00			
		-0.3			
Relative transaction size	δ_5	-0.10	-0.10 *		-0.07
		-1.3	*-1.6*		*-1.0*
Year of M&A	δ_6	0.00			
		-0.1			
Nobs	75				
Adj. R^2		-0.03	0.02	0.01	0.00
F-stat. (p-value)		0.83	0.28	0.45	0.58
DW-statistic		2.20	2.20	2.25	2.17

***, **, and * indicate statistical significance at the 1%, 5%, and 10% levels, respectively. t-statistics in italic; *CAB*: cumulative abnormal beta shift.

While overall the regression models are statistically insignificant, we find weak evidence for a couple of potential drivers of abnormal beta changes. First, regression model M2 suggests a potential negative relationship between emerging market transactions and systematic risk. The negative coefficient is significant at the 10% level. Second, while we did not find an impact of the relative transaction size in the risk event study, regression model M2 though suggests a potential negative relationship between relative transaction size and abnormal beta change. Again, the negative coefficient is significant at the 10% level. Contrary to our expectations, we do not find any significant coefficients for cross-border transactions, transaction value, nor transactions by the "big four".

Our second multivariate regression analysis for the rivals takes the form outlined in Equation 3.5:

$$CAB_{rival} = \alpha + \delta_1 * MM + \delta_2 * TV + \delta_3 * RS + \delta_4 * Yr \quad (3.5)$$

Geographical scope: The results of the risk event study suggested a positive relationship between rival *CAABs* and M&A in mature markets. We include a respective dummy variable *MM* in our regression analysis in order to account for mature market transactions and to verify the preliminary risk analysis.

Transaction value and relative transaction size: While we did not find significant proof to suggest a relationship between acquirer *CAABs* and the size of acquirer or target, we include both transaction value *(TV)* and relative transaction size *(RS)* in our regression analysis.

Transaction date: The risk event study suggested a change in rival *CAABs* over time. We include the year of the transaction announcement, *Yr*, in the regression analysis in order to account for a potential dependency.

Table 3.9 presents the results of four regression models (M1-M4) on the rival *CABs* for the [0;5] event window. Nonetheless, we have found significant rival *CAABs* for several event windows we decide to again perform our second multivariate analysis with respect to the [0;5] event window to stay consistent with the previous analysis. The regression models differ with regard to the independent variables included. Models M1 to M3 are statistically significant at the 5% or 10% level (based on F-statistics). Explanatory power is higher than for the acquirer analysis with adjusted R-squared ranging between 2% and 7%. Autocorrelation issues can be ruled out as the Durbin-Watson-statistics are close to two in all cases. The constant, alpha, is insignificant for all three models.

Table 3.9: Multivariate Regressions on Rival CABs

Dependent variable	*CAB* [0;5]				
Model		M1	M2	M3	M4
constant	α	-23.6	-0.02	-0.02	-0.02
		-1.1	*-0.6*	*-0.6*	*-0.6*
Independent variables					
Target mature market	δ_1	0.06			0.11 *
		1.1			*1.7*
Transaction value	δ_2	0.00 ***	0.00 ***		
		-3.0	*-2.6*		
Relative transaction size	δ_3	0.34 ***	0.34 ***	0.20 *	
		3.5	*3.8*	*1.8*	
Year of M&A	δ_4	0.01			
		1.1			
Nobs	96				
adj. R^2		0.06	0.07	0.04	0.02
F-stat. (p-value)		0.08 *	0.03 **	0.10 *	0.26
DW-statistic		1.74	1.71	1.67	1.75

***, **, and * indicate statistical significance at the 1%, 5%, and 10% levels, respectively. t-statistics in italic; *CAB*: cumulative abnormal beta shift.

The risk event study suggested a positive relationship between transactions in mature markets and rival *CAABs*. We do find evidence in the multivariate analysis to support such a relationship. But, evidence appears at low significance only testing the variable individually and the F-test does not approve the respective model M4 as generally significant. In addition, the multivariate analysis yields significant results with regard to the impact of transaction value and relative transaction size. The regression analysis suggests a negative impact of transaction value as indicated with negative t-statistics. Relative transaction size though shows a strongly positive effect on rival *CABs*. The corresponding coefficients are highly significant for two of three regression models. Finally, again contradicting the risk event study, the multivariate results do not suggest a change in *CABs* over time.

3.5. Conclusion

The objective of this study was to measure and analyze the abnormal short-term systematic risk shifts of acquirers and rivals in the brewing industry. For this purpose, a sample of 75 horizontal M&A transactions involving brewing companies between 1991 and 2010 was identified and analyzed using a combination of two approaches: the traditional event study methodology and the GARCH model based on Bollerslev (1986) and Engle (1982). Our results provide new insights into value implications of M&A transactions in the brewing industry through risk dynamics.

In order to provide a comprehensive review of short-term risk effects of M&A we analyzed a global dataset, which enabled us to specifically build and investigate relevant subsamples to determine key drivers of systematic risk. Given the dominance of the "big four" in the global brewing industry, we specifically analyzed abnormal risk as well as rival effects among them. In addition to the risk event study subsample analyses, the potential drivers were scrutinized by two multivariate regression models.

Our results indicate that on a total sample basis brewers do not experience significant abnormal risk shifts following the announcement of M&A transactions. These results are in line with long-term studies by Haugen and Langetieg (1975), Pettway and Yamada (1986), and Amihud, Dedong, and Saunders (2002) all of which provide evidence of no changes in systematic risk as a result of M&A announcements.

However, our results provide evidence for a number of drivers of systematic risk adjustments focusing on certain acquirer, target, or transaction characteristics within subsamples. Specifically, we find significantly negative acquirer abnormal risk shifts for cross-border transactions, in particular when targets are based in emerging market economies. These findings are in line with Kiymaz and Mukherjee (2001) and thus confirm the argument that diversification through cross-border transactions provides risk reduction and therefore value-enhancing opportunities. Moreover, we find evidence that suggests a negative relationship between systematic risk and transaction value.

Furthermore, we provide evidence that the "big four" experience significantly negative abnormal risk shifts when announcing a M&A transaction. At the same time, we find significantly positive abnormal risk shifts for close rivals missing out on a potential M&A opportunity, in particular when targets are based in mature markets where the beer market is generally fairly concentrated and thus M&A opportunities are limited. Moreover, our analysis suggests a positive relationship between rival companies' systematic risk and relative transaction size as well as a negative relationship with transaction value. Overall, our analysis on risk dynamics complements existing literature on short-term wealth effects, which generally assumes risk levels to be unaffected by M&A announcements and thus potentially misstates total shareholder value effects.

At the same time, it should be noted that there are also limitations to our study as our analysis is solely focused on the brewing industry. Due to the innovative and new risk event study methodology utilized there are no comparable results available. It could thus be questioned whether the determined results are truly industry-driven or general M&A-related effects.

Having identified the negative short-term risk implications of cross border M&A as well as determinant drivers, the question arises whether the short-term reductions in systematic risk are sustained in the long-run. So far there exists no empirical evidence analyzing the long-term implications of M&A on systematic risk of brewers. Hence, further empirical evidence should be provided by future research.

4. Bank Failures and Real Estate Investment Trust (REIT) Returns[10]

4.1. Introduction

Many consider REITs a separate asset class due to differences with common equities. REITs' risk-return behavior has been shown to be a combination of real estate and common stock market behavior that does not provide a definite affiliation to either one. As REITs are highly dependent on external financing this may possibly degrade their relative performance during times of high market uncertainty. However, Bredin, O'Reilly, and Stevenson (2007) have shown that REITs behave comparable to common equities when external monetary shocks, like monetary rate adjustments, occur. Hence, if for example, the price of liquidity rises this does not necessarily lead to a relative underperformance of REITs.

In our research we examine the case whether relative REIT performance changes when not the price but instead primarily the supply of liquidity declines due to high market uncertainty. We analyze the behavior of REITs relative to common equities in response to a surprise shock that stems from the inside of the system and occurs when market uncertainty is high. As Diamond and Rajan (2005) show bank failures represent such an internal shock to monetary markets. Because bank failures may be strongly contagious in integrated financial markets, a potential markets' phobia of system meltdown can cause a considerable decrease in market liquidity supply.

Bredin, O'Reilly, and Stevenson (2007) analyze the effect of monetary rate adjustments (as monetary shocks) on REIT markets. They state that REITs are also appropriate in representing general equity markets. They believe REITs to be the least biased in reflecting the equity markets' risk-return reactions to monetary rate changes since regulation requires a 90% earnings-payout ratio, hardly leaving REIT managers any room for applying potentially return biasing dividend strategies. Even though this is well defined research, we are curious whether REITs have similar behavior to common equities during times of high market uncertainty.

---

[10] I thank Jan-Willem Olliges for his major contribution to this study concerning the study structure. I am also grateful to Dirk Schiereck and Glenn Mueller for valuable comments and insights. Special thanks go to the conference participants at the annual meetings of the Academy of Economics and Finance 2011 in Jacksonville and of the American Real Estate Society 2011 in Seattle for valuable feedback.

Our hypotheses presume that a deviation of REIT return behavior to common equity returns critically depends on the type of shock to the market. Aiming for a shock to the market where this does not hold true we choose to use bank failures instead of monetary rate increases. While both types of shocks lead to liquidity reductions, the market finds itself in different situations regarding return prospects. We assume that monetary rate increases as external shocks occur when markets are confident about economic growth prospects, whereas bank failures as internal shocks usually occur at the peak and in the aftermath of a market crash when uncertainty about future prospects is still significant. While monetary rate increases represent a planned external shock to the market in times of high GDP growth and strong market outlooks, bank failures are quite different. During times of economic uncertainty and low or negative GDP growth, bank failure occurrences are unplanned and thus represent an internal shock that increases market uncertainty. As has been shown by Diamond and Rajan (2005) bank failures can be contagious and a consecutive shrinking in the common pool of liquidity might lead to liquidity shortages representing a monetary shock.

We analyze the impact of bank failures on U.S. REIT prices with respect to shifts in prevalent risk-return profiles. Using two approaches we first focus on the impact of 130 bank failure announcement dates between January 1999 and July 2010 on the all-REIT index and on ten REIT investment type subindices. In a second step, we consecutively scrutinize the impact of three singular large scale bank failures on 115 single REIT prices. The three failures that caught exceptional global attention were IndyMac Bank, Lehman Brothers, and Washington Mutual. For all events we analyze short window abnormal returns applying event study methodology and abnormal systematic risk shifts using an adapted risk event study. We are eager to determine whether bank failures have affected REITs' risk-return profile differently than other stocks.

We find positive abnormal returns averaging 0.93% for an all-REIT index around U.S. bank failure events relative to the S&P 500. Risk levels are stable on a total sample basis. However, these results are driven by REITs' reaction to bank failures during the recent financial crisis. In our analysis we control for two time periods: The *New Economy*-phase denotes the period from 1999-2004 and the *Real Estate Securitization*-phase denotes the consecutive period until 2010. During the *New Economy*-phase the all REIT index shows some evidence of a negative response to bank failures on both, the return and risk side. This observation is not statistically significant however. During the *Real Estate Securitization*-phase, which comprises the financial crisis, we find, e.g., a positive *CAAR* (median of 0.95%) for the event window [-5,+5 days] at the five percent significance level with no abnormal beta shifts. Controlling for capital investment type, Equity REITs uncover as the driver of positive abnormal returns observed in the previous analyses. Analyzing property focus of REITs we find primarily Retail, Residential, and Healthcare REITs to show value increasing *CAAR*-driven shifts in their risk-return profile.

In sum, the findings serve as strong support to our alternative hypothesis. Especially Equity REITs and respective property focused REITs that carry the least risk provide evidence for the safe haven theory regarding beneficial real estate valuation in times of crisis. On the contrary, Mortgage REITs, Diversified REITs and Industrial (Equiy) REITs show rather negative shifts in their risk-return profile. This indicates that investors do not consider investment in real estate backed loans and riskier direct real estate assets as comparable safe haven investments during turbulent times.

The paper is structured as follows: In part 4.2 we review the literature; in part 4.3 we explain our data selection and the empirical methodology; in part 4.4 we present our results; part 4.5 concludes.

4.2. Literature Review

4.2.1. Market Liquidity, Bank Failures, and REIT Behavior

Bredin, O'Reilly, and Stevenson (2007) investigate the influence of unanticipated changes in the U.S. monetary policy on Equity REITs. In contrast to previous studies which explore the influence of adjustments in monetary rate, their investigation focuses on unanticipated rate changes and their possible influence on REIT returns and volatility. Their results show a strong response in both first and second moments of REIT returns. The signs of the coefficients in relation to mean and volatility support the anticipation effect, i.e., negative for mean and positive for volatility. Also, they find evidence for a positive Friday effect in both the first and second moments of daily REIT returns.

Bernanke and Kuttner (2005) discuss the impact of changes in monetary policy on broader equity returns. Their findings indicate that the market reacts significantly to unanticipated monetary rate changes. Furthermore they document, that some effect of monetary policy on equity returns can be related to the revision in cash flows, but very little is directly attributable to changes in expected real interest rates. Their finding that the stock market reacts quite strongly to unexpected changes in monetary policy – an unanticipated raise of 25-basis-points typically leads to a decrease in stock prices on the order of 1% which is significant, but the question that is left unanswered is why stock prices respond as they do to monetary policy.

Diamond and Rajan (2005) discuss the implications of a banking crisis as a shock to the market. They find that these can shrink markets' common liquidity pool, potentially resulting in liquidity shortages. More critically, they find that bank failures could be contagious leading to a total meltdown of the system. Their objective is to find evidence on whether bank failures can start and propagate a systematic crisis because contagion stems from an underlying real general equilibrium problem. Therefore, Diamond and Rajan relate their study to policy perspectives, pointing out that simple confidence-building measures cannot be successful. Their paper concludes that banking crises can have significant impacts on market liquidity comparable to monetary rate adjustments.

Dell'Ariccia, Detragiache, and Rajan (2008) discuss the cause-effect relationship between a banking crisis and subsequently observable times of low credit and GDP growth. They show that banking crisis can externally hinder real activity, especially in those sectors that are more dependent on external financing. Due to the strict regulation of earnings distributions to shareholders, REITs cannot retain as many earnings as other market participants to fund future growth. Consequently, REITs are continuously dependent on external financing which should harm their performance in relation to common equities during credit crisis.

REIT prices do respond to real estate related news according to Ghosh, Guttery, and Sirmans (1998). Nevertheless, Giliberto (1990) found that the correlation between non-securitized real estate returns and REIT returns was close to zero. Ghosh, Miles, and Sirmans (1996) argue REITs are strongly influenced by stock market volatility in the short-term but their long-term return is derived from direct real estate growth. The interdependencies are never constant, however. The link between REITs, real estate, and other asset classes constantly changes as demonstrated by Clayton and MacKinnon (2001). Especially in a financial crisis, REIT returns might be influenced much more by market sentiment than in other times. The authors Basse and Friedrich (2009) reexamine the relationship between REITs and utility stocks in the current financial crisis and conclude that the crisis led to a structural change in this relationship. Their findings show that investments in REITs have become much more risky in times of crisis. Ambrose, Lee, and Peek (2007) analyze spillover effects across asset classes and find that since REITs have been added into three S&P general market indices they co-move more with common equities than their real estate fundamentals justify.

Mueller and Mueller (2003) highlight real estate's utility as an income producer and inflation hedge which could make REITs a preferred "safe haven" for investors in times of economic uncertainty. While they found such characteristics on the basis of data prior to 2002, we attempt to validate whether their findings still hold true for time periods thereafter and during financial crisis.

4.2.2. Contribution to the Literature and Hypotheses

Empirical evidence of the recent decade shows that market liquidity and its implications on real market behavior is of high interest to scholars in both financial economics and real estate economics. Therefore, we are eager to reexamine on the issue of how REITs perform relative to common equities during unique shocks to monetary markets. In the case of monetary rate increases, as an external shock increasing the price of liquidity, studies have found REITs show comparable risk-return behavior in size and direction to common equities. We present evidence that argues banking crises represent an internal type of a monetary shock that can result in a decrease of liquidity supply and may induce different effects. We believe analyzing bank failure events as an internal shock to monetary stability and its unique impact on REITs adds well to the literature.

Regarding behavioral expectations of REITs' return movements around bank failures, we may expect behavior similar to common equities, however. The occurrence of return behavior comparable to that of common equities would seem to be reasonable because bank failures may lead to a comparable decrease in liquidity as caused by monetary rate increases. REITs' continuous need for external financing may even constitute a further drawback as refinancing rates increase. This valuation effect on REITs through bank failures constitutes our null hypothesis.

The alternative hypothesis builds on the fact that bank failures occur during times of high market uncertainty in contrast to monetary rate increases that are triggered to prevent an overheating of the economy. Hence, the induced liquidity shocks are not comparable but differentiate strongly from each other. An increase in the liquidity's price due to monetary rate adjustments calms down the markets, whereas an unexpected decrease of liquidity supply due to bank failures may strongly increase market uncertainty. As a result bank failures could cause a relatively positive valuation effect for REITs because investors commonly interpret real estate as *safe haven* during economical crisis as shown by Mueller and Mueller (2003). Especially during a great financial crisis that causes strong market uncertainty, REITs may follow less short-term common equity behavior and move more in line with direct real estate valuation. Clayton and MacKinnon (2001) show that REIT returns are strongly driven by market sentiment during crisis but the question remains unanswered whether they suffer less than common equities. Consequently, the alternative hypothesis assumes that

REITs do not mirror common equities in the case of monetary shocks through bank failures as they do in the case of monetary shocks through monetary rate increases.

4.3. Data and Methodology

4.3.1. Data Selection and Descriptive Statistics

The sample of U.S. bank failure announcements is taken from the *Failed Bank List* provided by the Federal Deposit Insurance Corporation (FDIC). We use all U.S. bank failure dates between January 1999 and July 2010, thus capturing two economic cycles. The total sample of bank failures includes 130 event dates reflecting a total of 308 single bank failures totaling about USD 2.2 Trillion in assets. Figure 4.1 illustrates the total bank failure dates sample using a box plot. It shows the total assets at default-distribution per event date. Intuitively obvious, even to non-econometricians, the variable "assets at default" is not normally distributed. The sample is rightly skewed (skewness of 10.8) and characterized by fat tails (kurtosis of 119.7). Clear outliers in terms of total assets at default are the failures of Lehman Brothers (USD 1.6 Trillion), Washington Mutual (USD 307 Billion) and IndyMac Bank (USD 32 Billion).

Figure 4.1: Failed Banks' Total Assets at Default Distribution across Bank Failure Dates

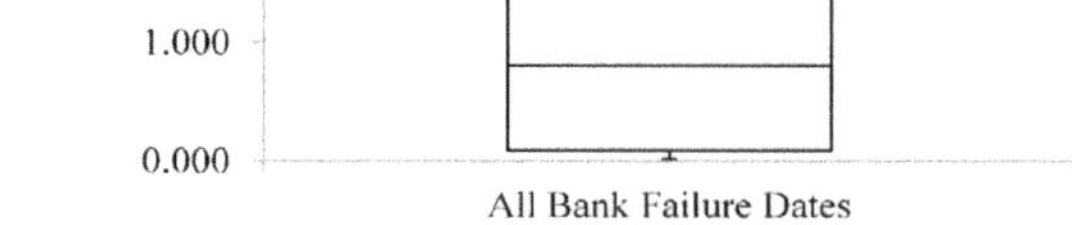

Comment: Number of bank failures: 308; number of bank failure dates: 130
Source: illustration by authors.

On average [median] USD 803 Million of defaulted total assets were recorded per event date. General inspection of our data shows that the FDIC announced 86.2% of the sample's bank failures on Fridays. Additionally, the announcements name certain financial services institutions that take over the failed institutions' operations as new owners (after the FDIC had covered a certain amount at default, the FDIC costs). Calendar effects and M&A within the financial services sector may thus have a biasing effect on our analysis. Therefore, we conduct robustness checks on both issues observing REITs' risk-return profile to obtain measures on the potential magnitude of these biases.

Frequently, multiple bank failures were announced per event date which explains lower average total assets at default-values at the bank level. Average [median] total assets at default per failed bank were USD 258 Million with an average USD 228 Million of deposits per bank. The FDIC had to cover a total of USD 73 Billion (on average USD 67 Million per bank)[11] with respect to failed banks in the sample. More descriptive statistics on the bank failure dates are shown in Table A4.1 in the appendix.

The stock market and REIT return data comes from Datastream. Respective return series are used from January 1, 1998 to July 31, 2010. Our REIT return data consists of four parts: First the FTSE NAREIT all-REIT index; second the FTSE NAREIT U.S. Equity REIT index, the U.S. Mortgage REIT index and the DS U.S. Diversified REIT index; third the seven property type U.S. Equity REIT indices (industrial, lodging, office, retail, residential, healthcare, and specialty), and fourth up to 117 single U.S. REITs. The single REITs are only used for our robustness analyses regarding the three individual bank failures and the three single regular bank M&As. The S&P 500 is used as our overall market return index. Table A4.2 in the appendix provides an overview of our REIT indices.

We apply two types of bank failure samples: Our total sample of U.S. bank failures as reported by the Federal Deposit Insurance Corporation is made up of 130 event dates between January 1999 and July 2010. The three largest failure events are also analyzed in a separate analysis. We apply standard event study methodology to measure short-term abnormal returns around bank failure dates. To obtain comparable results on a risk perspective, we use a beta-aligned event study approach. Determining daily REIT beta measures by applying a simple GARCH (1,1) model enables us to capture market adjusted abnormal beta shifts within short-term windows, equivalent to the standard event study.

Dividing our sample by the most recent economic cycles provides us with two subsamples: the *"New Economy: boom bust"* for the time period between January 1999 and December 2004 that contains 37 bank failure events, and the *"Real Estate Securitization: boom bust"* for the time period between January 2005 and July 2010 that contains 90 bank failure events (excl. the three largest

---

[11] FDIC statements do not always include information on FDIC costs at bank defaults. Thus, we present our given cost levels as minimum values.

bank failures). There were no bank failures in 2005 or 2006. Figure A4.1 in the appendix provides the sum of total bank assets defaulted per year. It shows that the magnitude of bank failures during the more recent period was a multiple of the earlier crisis and was not over in July 2010. It is not only the absolute amount of failures that rose, but also the average size and dispersion increased strongly because multiple failures on a single day became increasingly common. Figures A4.2 and A4.3 provide additional information on changes in bank failure characteristics on a yearly basis. We examine the three major outlier transactions of the recent financial crisis separately. These are the failures of IndyMac Bank (07/11/2008), Lehman Brothers (09/15/2008), and Washington Mutual (09/25/2008).

Four observations may show potential biases in the analyses:

First, at each specific bank failure date up to nine institutions declared bankruptcy at the same time. This is why we denote the events as bank failure dates instead of bank failures. We used the total value of defaults as a measure of potential impact, and not the number of failures per event, which may have a biasing effect on the analyses.

Second, due to the large amount of bank failures in short time periods the average time gap between two bank failure dates was 32 days. This could lead to overlap effects in event timing. We controlled this biasing issue by using the maximum event window size of [-5,+5 days] around the event date.

Third, 86.2% of bank failure dates were on a Friday. We control for this calendar effect bias by performing a separate analysis for all non-failure Fridays scrutinizing their effects on REITs' return moments.

Fourth, bank failures commonly result in bank M&A, because former competitive financial institutions continue the failed banks' operations and accounts. To identify any potential bias, we perform a robustness check on non-failure induced bank M&A (regular M&A).

4.3.2. Methodology of the Return and Risk Event Study

Our approach to the risk and return analysis of REITs regarding bank failures is twofold. For the return part we apply a standard event study methodology similar to MacKinlay (1997). We use a market model approach according to the Capital Asset Pricing Model (CAPM) by Sharpe (1964) and Lintner (1965). Performing simple OLS regressions from 280 to 30 days prior to each event we obtain case specific betas *(β)* as systematic risk estimators that enable us to forecast expected REIT returns dependent on actual market returns. Subtracting actually observed REIT returns around the event dates from expected REIT returns we obtain abnormal returns, *AR*. On a single event base the abnormal returns are cumulated *(CAR)* along differently sized event windows. Taking the equally weighted averages of the *CARs* with respect to the number of events we obtain the cumulated average abnormal returns, *CAAR*, for the event windows [0], [-1, +1], [-2,+2], [-5,+5], [0,+1], [0,+2], and [0,+5]. As per Massey (1951), the Kolmogorov-Smirnov test is used to determine whether the *CAARs* are based on normally distributed data. Then *CAAR* means are tested for difference from zero with an adjusted parametric t-test as developed by Boehmer, Musumeci, and Poulsen (1991) and a standard t-test. The medians are tested for significance applying the non-parametric Wilcoxon signed rank test.

For the risk part we apply an adapted event study methodology that is used to measure short-term shifts in beta that exceed or fall below expectations. The mathematical approach is equivalent to the standard event study. Our risk event study is composed of three basic steps: First, measuring daily REIT betas, second, estimating expected betas from before bank failure events, and third, subtracting measured betas from expected betas to obtain abnormal beta shifts, *AB*. Cumulating *ABs* within event windows and determining the arithmetic mean across events finally provides the cumulative average abnormal beta shifts, *CAAB*. Intuitively obvious to scholars of financial economics, measuring time-variant beta breaches the basic assumption of the CAPM stating that beta is stable over time. But, financial economists like Fabozzi and Francis (1978), Sunder (1980), Bos and Newbold (1984), Collins, Ledolter, and Rayburn (1987) and Jagannathan and Wang (1996) have argued against the assumption of beta stability and shown that beta is actually stochastic and thus conditional.

To determine time-varying volatility as an input parameter for daily beta measures, we apply the generalized autoregressive conditional heteroscedasticity (GARCH) approach. It is one of the most commonly applied techniques by researchers to observe changes in risk levels. The methodology was first introduced by Engle (1982), refined by Bollerslev (1986) and further advanced to a Multivariate GARCH (M-GARCH) by Bollerslev (1990). See Giannopoulos (1995) for an empirical implementation of GARCH models where he finds evidence that such volatility modeling techniques can be used not only to measure total risk but also systematic and unsystematic risk components. Brooks, Faff, and McKenzie (1998) investigate the forecasting accuracy of three techniques regarding time-variant beta: a M-GARCH, a time-varying beta market model suggested by Schwert and Seguin (1990), and the Kalman filter for a sample of returns on Australian industry portfolios. Furthermore, Faff, Hillier, and Hillier (2000) applied an M-GARCH approach to model time-varying beta in their study discussing alternative beta risk modeling techniques. They define an additional assumption of a constant correlation coefficient between the returns of an asset and respective market returns. The behavior of this correlation coefficient is core to our risk event study determining *CAABs*.

We apply a simple GARCH (1,1) model. It determines a specific asset's time-varying volatility measure as a weighted sum of a long-term average variance rate, Φ_i, the previous day return, $u_{i,t-1}$, and the previous day standard deviation, $\sigma_{i,t-1}$. The weighting factors γ_1, γ_2, and γ_3 are determined by applying the iterative Maximum Likelihood technique as discussed by Hull (2006).

$$\sigma_{it}^2 = \gamma_1 \Phi_i + \gamma_2 u_{i,t-1}^2 + \gamma_3 \sigma_{i,t-1}^2 \tag{4.1}$$

Obtaining beta measures as the next step we apply the M-GARCH model that observes the actual betas as a product of a constant correlation coefficient, ρ_{im}, between the specific REIT returns and the market returns (S&P 500 Composite), and the time-varying REIT standard deviation, σ_{it}, divided by the time-varying market standard deviation, σ_{mt}.

$$\beta_{it} = \frac{\rho_{im}\sigma_{it}}{\sigma_{mt}} \tag{4.2}$$

Finishing the empirical groundwork, the following data is available: Daily volatility measures for all single REITs, REIT indices and the S&P 500 Composite; daily correlation coefficients for all single REITs as well as REIT indices with the S&P 500 Composite, based on a one year lag structure; and daily beta measures for all single REITs as well as REIT indices in relation to the S&P 500 Composite.

Equivalent to beta for the standard return event study we utilize the correlation coefficient 30 days prior to the event for our systematic risk estimates, $E(\beta_{it})$. REIT and market volatility measures remain time variant as the market return does in the standard event study. The abnormal beta shift, *AB*, is then determined as the difference between expected and measured beta from five days before to five days after the event.

$$AB_{it} = E(\beta_{it}) - \beta_{it} = \frac{\rho_{im}\sigma_{it}}{\sigma_{mt}} - \beta_{it} \tag{4.3}$$

Cumulating *ABs* along the differently sized event windows [0], [-1,+1], [-2, +2], [-5,+5], [0,+1], [0,+2], and [0,+5] we obtain the cumulative abnormal beta shifts, *CAB*. The arithmetic mean of *CABs* across the number of events, *n*, per analysis returns the cumulative average abnormal beta shifts, *CAAB*. Again, the Kolmogorov-Smirnov test is used to determine whether these *CAABs* are based on normally distributed *CABs*. The non-parametric Wilcoxon signed rank test and the parametric t-test are applied to test the *CAABs* for statistical difference to zero.

4.4. Results

4.4.1. Bank Failures' Effect on the All-REIT Index

We start with a total sample analysis, measuring the effect of all bank failure dates in our sample on REITs' risk-return profiles. Before, we derive directional and intensity expectations on the risk and return behavior as a consolidated adjustment in the prevalent REIT risk-return profile. We also identify bank failure characteristics as potential drivers and scrutinize likely drivers on the REIT side.

The following factors of bank failure dates are examined: Event date on the time line, total bank assets at default, total deposits at default, total FDIC costs at default, and number of banks failed per event date. Descriptive statistics of the bank failure dates sample show that there is a high degree of serial correlation among all of these characteristics leading us to use the event date on time line characteristic representatively for a further analysis step. Our rationale is that over time consolidation in the financial services industry drove growing asset values as shown by Beitel, Schiereck, and Wahrenburg (2004) leading also to higher deposits and linked FDIC costs at default. Furthermore, the two consecutive crises under analysis may be directional determinants of REIT return behavior. Asset and deposit values or FDIC costs are rather amplifiers of REITs' reaction intensity.

REITs are analyzed using the following characteristics: Capital investment type focus and property type focus. On the capital side we compare the risk-return behavior of Equity REITs with Mortgage REITs and Diversified REITs. For Equity REITs we then also compare the risk-return behavior of the seven major property sector focused REITs.

To cross-check the empirical findings we use two robustness tests: First, we examine non-failure Fridays' effects on REITs and non-failure induced bank M&As' effects on REITs. Second, we investigate the effects of the three major outlier bank failures on 115 single REITs. We again compare these to the implications of three single regular bank M&A events on the same single REITs.

Our first analysis considers the total bank failure dates sample including 130 event dates and observes their effect on the U.S. REIT market by measuring changes in the FTSE NAREIT index's risk-return profile. Compared to the case of external monetary shocks in form of unexpected monetary rate increases, we expect bank failures as internal monetary shocks to have the same effect on

REIT returns and beta risk as we would see on common equities. Thus, our null hypothesis assumes there should be no significant abnormal returns or abnormal risk shifts in the REIT risk-return profile versus the S&P 500 Composite. However, we may also argument for a positive shift in REITs' risk-return profile due to the safe haven theory. Our alternative hypothesis therefore constitutes that REITs participate in the beneficial real estate valuation during times of high market uncertainty.

As Table 4.1 shows significantly positive abnormal returns are found for three event windows (mean values displayed): 0.44% at event date, 0.93% for the symmetric [-5,+5] window, and 0.69% for the asymmetric [0,+5] window. Regarding risk, there are no abnormal changes for beta observable. Thus, from total sample analysis we can infer that bank failures induce a positive relative return shift for REITs compared to common equities, while REIT systematic risk stays relatively stable. This results in a positive valuation effect for North American REITs due to U.S. bank failures in comparison to common equity markets and thus supports our alternative hypothesis at an overall level.

Table 4.1: Risk/Return Event Study: All Bank Failure Dates

[-x; +y]	Mean		Median		KS-test	p	Wx-test	p	Z-stat.	p	t-test	p	N
CAAR													
0 0	0.44%	**	0.06%	**	0.19	0.00	-1.62	0.11	2.07	0.04	2.10	0.04	130
1 1	0.14%		0.05%		0.12	0.04	-0.15	0.88	0.56	0.57	0.66	0.51	130
5 5	0.93%	***	0.75%	***	0.10	0.17	-2.53	0.01	2.52	0.01	2.23	0.03	130
0 1	0.29%		0.31%		0.12	0.03	-1.50	0.13	1.40	0.16	1.52	0.13	130
0 5	0.69%	**	0.55%	**	0.09	0.19	-2.22	0.03	2.34	0.02	2.24	0.03	130
CAAB													
0 0	0.00		0.00		0.10	0.15	-0.52	0.61	-	-	0.78	0.44	130
1 1	0.01		0.00		0.10	0.13	-0.43	0.66	-	-	0.75	0.45	130
5 5	0.02		0.00		0.10	0.14	-0.24	0.81	-	-	0.71	0.48	130
0 1	0.00		0.00		0.10	0.14	-0.46	0.65	-	-	0.74	0.46	130
0 5	0.01		0.00		0.09	0.19	-0.25	0.80	-	-	0.61	0.54	130

***, **, and * indicate statistical significance at the 1%, 5%, and 10% levels, respectively. *CAAR*: cumulative average abnormal return, *CAAB*: cumulative average abnormal beta shift, KS test: Kolmogorov-Smirnov test - p-value ≤ 0.05 indicates normal distribution at the 5% significance level, Wx-test: Wilcoxon signed rank test as non-parametric median test, Z-stat.: Boehmer test (parametric), t-test: parametric mean test.

The first subsample analysis addresses potentially diverging implications of bank failures on REITs due to different time periods. Dividing the sample into two economic cycles yields the "New Economy: boom bust" for the time period January 1999 to December 2004 that contains 37 bank failure dates, and the "Real Estate Securitization: boom bust" for the time period January 2005 to July 2010 that contains 90 bank failure dates. The three largest single bank failures in terms of assets at default are excluded from this analysis. Compared to all other bank failure dates the bankruptcies of IndyMac Bank, Lehman Brothers, and Washington Mutual represent clear outliers in terms of asset size and could strongly impact second period results compared to first period results. In a later analysis these three bank failures are analyzed separately.

The issue of serial correlation between bank failure dates in time, asset, and deposit values (increasing FDIC costs) constitutes the foundation of our risk-return behavior expectations for this subsample analysis. Assets, deposits and FDIC costs are on average four to seven times higher during the *Real Estate Securitization*-phase compared to the *New Economy*-phase. Thus, we expect effects of higher magnitude and enhanced significance in the latter phase. Since the previous analysis did not return expected results but showed positive abnormal returns for REITs, we expect a repetition of these results in terms of positive wealth effects for both time periods.

On the return side the NAREIT index shows one significantly negative response to bank failures during the *New Economy*-phase. A single negative abnormal return, median *CAAR* of -0.34%, for the event window [-1,+1] is significant at the ten percent level. Additionally, there are two negative cumulative average abnormal beta shifts at the ten percent significance level which provide some evidence for an unexpected systematic risk decline. Table 4.2 displays the results of the respective analysis.

During the *Real Estate Securitization*-phase our original finding is supported. For the event date we find a *CAAR* of 0.69% at a five percent signifi-cance level and for the event window [-5,+5] we find a *CAAR* of 1.18% at the five percent significance level. Two further event windows also indicate the po-sitive abnormal returns. But, we do not find any respective abnormal beta shifts, however.

Overall, REITs' risk-return profile shows several short-term shifts due to bank failure events during the differently characterized economic cycles. The distinct time periods did cause different REIT behaviors, signaling that bank failures' effects on REITs are not independent of time. The weak indication of an abnormal decline in systematic risk during the *New Economy*-period is probably offset by the negative abnormal returns at comparably low significance. Thus, there is basically no adjustment in risk-return profile throughout the first time period. During the second period, comprising also the financial crisis, strong evidence of positive abnormal returns clearly leads to higher REIT valuations. In addition we can report, that second period results clearly drive our total sample risk-return behavior observations.

Table 4.2: Risk/Return Event Study: All Bank Failure Dates and Time Period Subsamples

	[-x; +y]	Mean		Median		KS-test	p	Wx-test	p	t-test[1]	p	n
New Economy: boom bust (1999-2004)												
	0 0	-0.08%		-0.09%		0.13	0.52	-1.23	0.22	-1.27	0.20	37
	1 1	-0.27%	**	-0.34%	*	0.10	0.83	-1.79	0.07	-1.98	0.05	37
CAAR	5 5	0.24%		0.15%		0.08	0.94	-0.54	0.59	0.47	0.64	37
	0 1	-0.17%		-0.17%		0.09	0.89	-1.15	0.25	-1.34	0.18	37
	0 5	-0.09%		-0.16%		0.07	0.99	-0.16	0.87	-0.34	0.74	37
	0 0	-0.01		-0.01		0.12	0.59	-1.49	0.14	-1.51	0.14	37
	1 1	-0.02		-0.03		0.12	0.62	-1.49	0.14	-1.49	0.15	37
CAAB	5 5	-0.07		-0.11	*	0.09	0.87	-1.76	0.08	-1.60	0.12	37
	0 1	-0.01		-0.02		0.12	0.61	-1.47	0.14	-1.48	0.15	37
	0 5	-0.04		-0.06	*	0.09	0.86	-1.62	0.10	-1.55	0.13	37
Real Estate Securitization: boom bust (2005-2010)												
	0 0	0.69%	**	0.28%	**	0.15	0.02	-2.32	0.02	2.42	0.02	90
	1 1	0.28%		0.31%		0.11	0.24	-0.63	0.53	1.22	0.22	90
CAAR	5 5	1.18%	***	0.95%	**	0.08	0.64	-2.32	0.02	2.45	0.01	90
	0 1	0.47%	**	0.56%	**	0.13	0.10	-1.95	0.05	1.93	0.05	90
	0 5	0.98%	***	0.92%	**	0.10	0.35	-2.26	0.02	2.66	0.01	90
	0 0	0.01		0.00		0.13	0.09	-1.25	0.21	1.36	0.18	90
	1 1	0.01		0.00		0.13	0.07	-1.16	0.25	1.32	0.19	90
CAAB	5 5	0.05		0.01		0.12	0.13	-1.01	0.31	1.30	0.20	90
	0 1	0.01		0.00		0.13	0.07	-1.17	0.24	1.31	0.19	90
	0 5	0.03		0.01		0.11	0.18	-0.97	0.33	1.22	0.23	90

***, **, and * indicate statistical significance at the 1%, 5%, and 10% levels, respectively. *CAAR*: cumulative average abnormal return, *CAAB*: cumulative average abnormal beta shift, KS test: Kolmogorov-Smirnov test - p-value ≤ 0.05 indicates normal distribution at the 5% significance level, Wx-test: Wilcoxon signed rank test as non-parametric median test, [1]t-test: Parametric mean tests: Boehmer test for *CAARs*; Student's t-test for *CAABs*.

4.4.2. Bank Failures' Effect on Focused REIT Indices

The second subsample analysis concentrates on distinct REIT types and we hope to find some determinants that can add explanatory value to the results observed in the total sample analysis. We perform return and risk event studies on the Equity REIT index versus the Mortgage REIT index and the Diversified REIT index. Subsequently, we analyze seven different property focused Equity REIT types: industrial, office, retail, residential, healthcare, lodging, and specialized.

We start by comparing Equity REITs versus Mortgage REITs and Diversified REITs. While Equity REITs directly own equity stakes in properties, Mortgage REITs invest in loans backed by real estate and their returns are affected more by interest rates. This indirect real estate investment could cause negative valuation effects during crises. We expect a positive valuation effect for Equity REITs based on our total sample results, which would support the *safe haven* theory regarding beneficial real estate valuation during turbulent times. The *safe haven* theory should apply less in the case of Mortgage REITs as these invest indirectly in real estate only. In the case of Mortgage REITs, there is no reason to expect returns above or risk levels below those of common equity markets. Diversified REITs are expected to show an adjustment of their risk-return profile that lies between the former two REIT types. For previous evidence on the transmission of shocks to these three capital investment REIT types see also Payne and Mohammadi (2004) who yet document that shocks are transmitted rapidly across REIT markets.

Table 4.3: Risk/Return Event Study: REIT-type Subsamples (I/III)

		[-x; +y]	Mean		Median		KS-test	p	Wx-test	p	t-test[1]	p	n
Equity	*CAAR*	0 0	0.55%	**	0.20%	**	0.17	0.00	-2.10	0.04	2.33	0.02	111
		1 1	0.28%		0.34%		0.12	0.07	-0.94	0.35	1.49	0.14	111
		5 5	1.06%	***	0.97%	**	0.14	0.03	-2.13	0.03	2.77	0.01	111
		0 1	0.42%	**	0.49%	***	0.09	0.27	-2.69	0.01	2.14	0.03	111
		0 5	0.78%	***	0.57%	**	0.10	0.21	-2.21	0.03	2.44	0.01	111
	CAAB	0 0	0.00		0.00		0.12	0.08	-0.82	0.41	1.17	0.25	111
		1 1	0.01		0.00		0.13	0.05	-0.74	0.46	1.12	0.27	111
		5 5	0.04		0.00		0.12	0.08	-0.48	0.63	1.04	0.30	111
		0 1	0.01		0.00		0.13	0.05	-0.72	0.47	1.12	0.27	111
		0 5	0.02		0.00		0.12	0.07	-0.57	0.57	0.96	0.34	111
Mortgage	*CAAR*	0 0	0.43%		0.12%		0.16	0.01	-1.59	0.11	1.59	0.11	111
		1 1	-0.06%		0.11%		0.19	0.00	-0.46	0.64	-0.17	0.86	111
		5 5	-0.21%		-0.84%		0.10	0.18	-0.16	0.87	-0.22	0.82	111
		0 1	0.21%		0.25%		0.12	0.08	-1.32	0.19	0.75	0.45	111
		0 5	-0.27%		-0.22%		0.06	0.74	-0.36	0.72	-0.60	0.55	111
	CAAB	0 0	0.01		0.01	**	0.16	0.01	-1.93	0.05	0.86	0.39	111
		1 1	0.01		0.02	*	0.16	0.00	-1.87	0.06	0.76	0.45	111
		5 5	0.05		0.07	**	0.16	0.00	-1.97	0.05	0.82	0.42	111
		0 1	0.01		0.01	**	0.16	0.01	-1.93	0.05	0.84	0.40	111
		0 5	0.03		0.03	*	0.16	0.01	-1.78	0.08	0.74	0.46	111
Diversified	*CAAR*	0 0	-0.18%		-0.09%		0.11	0.09	-0.95	0.34	-1.21	0.22	130
		1 1	0.10%		-0.22%		0.08	0.29	-0.15	0.88	0.35	0.73	130
		5 5	0.68%		0.63%		0.07	0.57	-1.31	0.19	1.11	0.27	130
		0 1	-0.09%		-0.17%		0.07	0.53	-0.43	0.67	-0.62	0.54	130
		0 5	0.30%		0.31%		0.06	0.65	-0.56	0.58	0.47	0.64	130
	CAAB	0 0	0.01	**	0.01	**	0.10	0.14	-2.14	0.03	1.95	0.05	130
		1 1	0.02	*	0.02	**	0.08	0.31	-2.12	0.03	1.86	0.07	130
		5 5	0.07	**	0.08	**	0.09	0.22	-2.24	0.03	1.95	0.05	130
		0 1	0.01	**	0.02	**	0.09	0.21	-2.19	0.03	1.97	0.05	130
		0 5	0.05	**	0.03	**	0.09	0.24	-2.37	0.02	2.12	0.04	130

***, **, and * indicate statistical significance at the 1%, 5%, and 10% levels, respectively. *CAAR*: cumulative average abnormal return, *CAAB*: cumulative average abnormal beta shift, KS test: Kolmogorov-Smirnov test - p-value ≤ 0.05 indicates normal distribution at the 5% significance level, Wx-test: Wilcoxon signed rank test as non-parametric median test, [1]t-test: Parametric mean tests: Boehmer test for *CAARs*; Student's t-test for *CAABs*.

We find that the general Equity REIT index shows a strong positive return effect at high significance. Again, there is no respective risk shift. Mortgage REITs do not show any significant return shifts different from the common equity market. They do show a significant increase in abnormal systematic risk, however. Against expectations Diversified REITs behavior does not lie between Equity and Mortgage REITs but exceeds Mortgage REITs beta risk increase. Table 4.3 provides the respective result statistics.

In sum, the findings confirm our expectations only in parts. Equity REITs confirm expectations with their positive wealth effect and thus show that their behavior again drives total sample results. Mortgage REITs actually underperform common equities as their systematic risk level increases showing significance of all displayed median values. The fact that Diversified REITs show significance additionally in all mean abnormal beta increases lets us infer that their risk increase might be even stronger. Hence, Diversified REITs are evaluated most critically by investors during times of high market uncertainty, which may also be due to the fact that their investment strategy is not as transparent.

For property sector REITs we now examine the value increasing to value decreasing effects of expected risk-return profile behavior. Thereby, we build on previous findings by Payne (2006) who extended his earlier research on the transmission of shocks by an examination of REIT subsectors. Analyzing a comparable set of REIT subsectors as we do, he identifies that Lodging REITs show the largest initial response to shocks, whereas Residential REITs are affected the least. Furthermore, Ro and Ziobrowski (2011) may be considered for a performance study of property-focused REITs (all subsectors individually) versus Diversified REITs. We formulate the following expectations:

Industrial REITs: Investing in warehouse buildings, their steady cash flows are secured by long-term leases in multiuser properties which should make them less risky than other REIT types. Thus, we expect a less profound decline during crises implying a relatively positive valuation compared to common equities.

Lodging REITs: They invest in hotel properties. Hotels cater to business and leisure travelers to generate their cash flows. Travelling represents generally one among the first expense categories affected by budget restrictions during crises but certainly also rises strongly in boom times. Thus, Lodging REITs should earn higher returns at higher risk levels. Bank failures as symptom of continued market uncertainty are thus expected to rather decrease Lodging REITs value relative to common equities. See also Jackson (2009) whose findings suggest that Lodging REITs are less resilient as other property-focused REIT types.

Office REITs: Investing in, e.g., commercial downtown properties, they obtain their income on the basis of medium-term leases securing cash flows through short recessions but may suffer intensively from long periods of low GDP growth. Their property location close to urban centers should devalue relatively little during recessions due to high marketability and potential multi-functional and multi-tenant use. Zietz, Sirmans, and Friday (2003) provide a relevant performance analysis on more traditional REIT subsectors as, e.g., Office and Retail REITs.

Specialized REITs: Investing in smaller property segments such as public storage or timber, they rely on unique demand segments for their cash flows. Thus, Specialized REITs may operate at either low or high risk levels and experience either positive or negative valuation effects in relation to common equities.

Retail REITs: Investing in shopping centers and malls, their cash flow depends on household spending. As bank failures do not directly impact households, spending levels should not be decreasing significantly due to this type of event. Retail REITs are thus expected to increase their performance relative to common equities.

Residential REITs: Investing in rental apartment housing, they should undergo comparable risks to Retail REITs. Potentially, their investment focus limits them to lower income households compared to Retail REITs who also receive sales from wealthier segments of the population that reside in private housing. However, long-term rental arrangements should benefit their valuation during crisis. Newell and Fischer (2009) provide recent evidence on the relative performance of Residential REITs in REIT portfolios before and after the 2007 subprime and financial crisis. They find that this asset class benefits as home ownership decreases while people move into rental accommodations which receive special attention by institutional investors.

Healthcare REITs: Investing in seniors housing and medical offices, they receive income that depends mainly on insurance and retirement payments that should be recession resistant. Therefore, bank failures should not impact Healthcare REITs negatively and their values may actually strengthen during recessions. Newell and Peng (2006) should be considered for further evidence on the performance of rather non-traditional REIT subsectors such as healthcare, specialty, and, e.g., self-storage.

Table 4.4 shows for the REIT types industrial, lodging, office, and specialized that they underperform their REIT type peers retail, residential and healthcare. Respective figures for the latter REIT types can be found in Table 4.5. Industrial REITs show hardly positive abnormal returns but strong abnormal beta increases leading to a negative wealth effect. This behavior is clearly below expectations. Lodging REITs show only few positive abnormal returns but no risk shifts which leaves their value basically unchanged. By this behavior Lodging REITs still perform above expectations. Office REITs and Specialized REITs show comparable positive abnormal returns in several event windows at maximum five percent significance. Since our expectations constituted rather zero value adjustments, both REIT types performed above expectations.

Retail REITs, Residential REITs, and Healthcare REITs all show strong positive abnormal returns up the one percent significance level. Their behavior confirms expectations as we expected these REIT types to be least risky and thus profit most from increased market uncertainty. In fact the REIT property type analysis further supports the *save haven* theory as Equity REITs with low risk property focus performed best among all Equity REITs with property focus and gained strong relative value compared to common equities.

Table 4.4: Risk/Return Event Study: REIT-type Subsamples (II/III)

		[-x; +y]	Mean		Median		KS-test	p	Wx-test	p	t-test[1]	p	n
Industrial	*CAAR*	0 0	0.82%	*	0.57%	*	0.16	0.02	-2.24	0.03	1.81	0.07	92
		1 1	0.17%		0.45%		0.15	0.04	-0.64	0.52	0.25	0.80	92
		5 5	0.35%		0.26%		0.14	0.06	-0.42	0.67	0.27	0.79	92
		0 1	0.66%		0.48%		0.18	0.00	-1.61	0.11	1.19	0.23	92
		0 5	1.07%		0.57%		0.18	0.00	-0.75	0.45	1.28	0.20	92
	CAAB	0 0	0.01	**	0.01	**	0.11	0.18	-2.39	0.02	2.20	0.03	92
		1 1	0.04	**	0.03	**	0.11	0.18	-2.29	0.02	2.22	0.03	92
		5 5	0.14	**	0.13	**	0.10	0.28	-2.27	0.02	2.26	0.03	92
		0 1	0.03	**	0.02	**	0.11	0.17	-2.31	0.02	2.23	0.03	92
		0 5	0.08	**	0.08	**	0.11	0.23	-2.33	0.02	2.32	0.02	92
Lodging	*CAAR*	0 0	0.54%		0.20%		0.15	0.04	-1.18	0.24	1.54	0.12	92
		1 1	-0.15%		-0.29%		0.10	0.26	-0.71	0.48	-0.17	0.86	92
		5 5	1.22%	*	1.07%		0.06	0.90	-1.48	0.14	1.91	0.06	92
		0 1	0.20%		0.27%		0.15	0.02	-0.29	0.77	0.56	0.58	92
		0 5	1.26%	**	0.97%	*	0.08	0.52	-1.84	0.07	2.23	0.03	92
	CAAB	0 0	0.01		0.00		0.11	0.18	-1.04	0.30	1.16	0.25	92
		1 1	0.02		0.00		0.11	0.20	-1.03	0.30	1.06	0.29	92
		5 5	0.05		0.00		0.11	0.19	-1.18	0.24	1.06	0.29	92
		0 1	0.01		0.00		0.10	0.28	-1.01	0.31	1.07	0.29	92
		0 5	0.03		0.00		0.11	0.23	-1.05	0.29	0.98	0.33	92
Office	*CAAR*	0 0	0.62%	**	0.23%	**	0.14	0.04	-1.93	0.05	1.94	0.05	92
		1 1	0.29%		0.16%		0.11	0.23	-0.94	0.35	1.12	0.26	92
		5 5	1.13%	**	0.45%	*	0.11	0.18	-1.80	0.07	2.08	0.04	92
		0 1	0.40%		0.11%		0.11	0.22	-1.11	0.27	1.52	0.13	92
		0 5	0.96%	**	1.00%	*	0.12	0.16	-1.64	0.10	2.27	0.02	92
	CAAB	0 0	0.01		0.00		0.12	0.12	-1.16	0.25	1.57	0.12	92
		1 1	0.02		0.00		0.13	0.09	-1.09	0.28	1.55	0.13	92
		5 5	0.07		0.00		0.12	0.15	-1.07	0.29	1.53	0.13	92
		0 1	0.01		0.00		0.13	0.08	-1.08	0.28	1.56	0.12	92
		0 5	0.04		0.00		0.12	0.13	-0.92	0.36	1.46	0.15	92
Specialized	*CAAR*	0 0	0.39%	**	0.14%	*	0.17	0.00	-1.80	0.07	1.94	0.05	130
		1 1	0.18%		0.03%		0.11	0.09	-0.40	0.69	0.89	0.37	130
		5 5	0.68%	**	0.75%	**	0.08	0.30	-2.22	0.03	2.42	0.02	130
		0 1	0.34%	*	0.21%	*	0.12	0.05	-1.73	0.08	1.63	0.10	130
		0 5	0.47%	*	0.05%		0.07	0.48	-1.18	0.24	1.85	0.06	130
	CAAB	0 0	0.00		0.00		0.13	0.02	-1.36	0.18	0.68	0.50	130
		1 1	0.00		0.01		0.13	0.02	-1.11	0.27	0.56	0.58	130
		5 5	0.02		0.03		0.12	0.04	-1.05	0.29	0.51	0.61	130
		0 1	0.00		0.01		0.13	0.02	-1.20	0.23	0.56	0.57	130
		0 5	0.01		0.02		0.14	0.01	-1.26	0.21	0.50	0.62	130

***, **, and * indicate statistical significance at the 1%, 5%, and 10% levels, respectively. *CAAR*: cumulative average abnormal return, *CAAB*: cumulative average abnormal beta shift, KS test: Kolmogorov-Smirnov test - p-value ≤ 0.05 indicates normal distribution at the 5% significance level, Wx-test: Wilcoxon signed rank test as non-parametric median test, [1]t-test: Parametric mean tests: Boehmer test for *CAARs*; Student's t-test for *CAABs*.

Table 4.5: Risk/Return Event Study: REIT-type Subsamples (III/III)

		[-x; +y]	Mean		Median		KS-test	p	Wx-test	p	t-test[1]	p	n
Retail	*CAAR*	0 0	0.57%	*	0.25%	**	0.15	0.03	-2.16	0.03	1.88	0.06	92
		1 1	0.38%		0.13%		0.09	0.46	-0.88	0.38	1.45	0.15	92
		5 5	1.42%	***	1.48%	**	0.10	0.30	-2.20	0.03	2.50	0.01	92
		0 1	0.37%		0.67%	*	0.12	0.16	-1.64	0.10	1.58	0.11	92
		0 5	1.06%	***	1.17%	**	0.11	0.17	-2.13	0.03	2.49	0.01	92
	CAAB	0 0	0.01		0.00		0.12	0.16	-0.83	0.40	1.11	0.27	92
		1 1	0.01		0.00		0.12	0.15	-0.78	0.43	1.12	0.27	92
		5 5	0.05		-0.01		0.13	0.09	-0.73	0.47	1.09	0.28	92
		0 1	0.01		0.00		0.12	0.15	-0.83	0.40	1.11	0.27	92
		0 5	0.03		-0.01		0.10	0.26	-0.62	0.54	0.99	0.32	92
Residential	*CAAR*	0 0	0.81%	***	0.37%	***	0.15	0.02	-2.39	0.02	2.67	0.01	92
		1 1	0.29%		0.27%		0.11	0.20	-0.55	0.58	1.27	0.20	92
		5 5	1.35%	**	1.68%	**	0.07	0.76	-2.22	0.03	2.37	0.02	92
		0 1	0.54%	**	0.40%	**	0.11	0.17	-2.18	0.03	2.25	0.02	92
		0 5	1.06%	***	0.60%	**	0.07	0.66	-2.25	0.02	2.57	0.01	92
	CAAB	0 0	0.00		0.00		0.14	0.06	-1.21	0.23	0.76	0.45	92
		1 1	0.01		0.01		0.13	0.09	-1.15	0.25	0.73	0.47	92
		5 5	0.04		0.03		0.13	0.08	-1.23	0.22	0.73	0.47	92
		0 1	0.01		0.00		0.13	0.07	-1.18	0.24	0.73	0.47	92
		0 5	0.02		0.00		0.11	0.21	-1.13	0.26	0.69	0.49	92
Healthcare	*CAAR*	0 0	0.75%	***	0.45%	**	0.13	0.08	-2.39	0.02	2.64	0.01	89
		1 1	0.17%		0.41%		0.07	0.76	-0.82	0.41	0.92	0.36	89
		5 5	0.70%		0.41%		0.10	0.32	-1.35	0.18	1.51	0.13	89
		0 1	0.49%	*	0.57%	**	0.11	0.19	-2.11	0.03	1.84	0.07	89
		0 5	0.76%	**	0.78%		0.05	0.96	-1.31	0.19	1.94	0.05	89
	CAAB	0 0	0.00		0.00		0.17	0.01	-0.13	0.90	-0.67	0.51	89
		1 1	-0.01		0.00		0.17	0.01	-0.07	0.94	-0.76	0.45	89
		5 5	-0.04		0.00		0.16	0.02	0.00	1.00	-0.88	0.38	89
		0 1	-0.01		0.00		0.18	0.01	-0.08	0.94	-0.75	0.45	89
		0 5	-0.02		0.00		0.18	0.01	-0.01	0.99	-0.90	0.37	89

***, **, and * indicate statistical significance at the 1%, 5%, and 10% levels, respectively. *CAAR*: cumulative average abnormal return, *CAAB*: cumulative average abnormal beta shift, KS test: Kolmogorov-Smirnov test - p-value ≤ 0.05 indicates normal distribution at the 5% significance level, Wx-test: Wilcoxon signed rank test as non-parametric median test, [1]t-test: Parametric mean tests: Boehmer test for *CAARs*; Student's t-test for *CAABs*.

4.4.3. Robustness Analysis regarding Potential Biases

Next, we discuss potential biasing factors regarding our obtained findings and perform analyses that control these factors. 86.2% of all bank failure dates occurred on a Friday and bank failures commonly result in bank M&A, so that accounts and operations can continue. Thus, we control for these variables and perform robustness checks on both issues: An analysis of non-failure Fridays' impact on REITs between 1999 and 2010 and an analysis of non-failure induced bank M&A activity's impact on REITs between 1999 and 2010. Scrutinizing these potential biases we find evidence on a marginally positive risk impact of the Friday effect on REITs (*CAAB* of up to 0.16 at a ten percent significance level). The robustness check on non-failure induced bank M&A activity returns positive *CAARs*, e.g., of 0.43% at a five percent significance level and positive *CAABs* of up to 0.07 at a ten percent significance level. Table 4.6 depicts the results.

Evaluating these results, we can infer that the marginally positive *CAABs* in the case of non-failure Fridays and regular bank M&A do not represent a core abnormal risk driver since we observed only few risk increases in the previous analyses. There is also no particular evidence that the Friday effect could have produced any negative beta shifts. Regarding the positive return effect of bank M&A, we cannot ignore some bias on our general results and will thus consider this issue in our final conclusion.

Table 4.6: Robustness Analysis on All-REIT Index Behavior

	[-x; +y]	Mean	Median	KS-test	p	Wx-ts.	p	Z-stat.	p	t-test	p	n
Non-failure Fridays												
CAAR	0 0	-0.16%	-0.06%	0.13	0.35	-0.40	0.69	-0.51	0.61	-0.44	0.66	46
	1 1	-0.75%	-0.23%	0.13	0.38	-0.96	0.34	-1.25	0.21	-1.26	0.21	46
	5 5	0.10%	-0.89%	0.15	0.23	-0.50	0.62	0.19	0.85	0.10	0.92	46
	0 1	-0.75%	-0.41%	0.14	0.29	-0.89	0.37	-1.26	0.21	-1.25	0.22	46
	0 5	-0.25%	0.16%	0.14	0.28	-0.07	0.94	-0.22	0.82	-0.31	0.76	46
CAAB	0 0	0.01	0.01 *	0.13	0.42	-1.69	0.09	-	-	1.65	0.11	46
	1 1	0.02	0.04 *	0.14	0.30	-1.71	0.09	-	-	1.62	0.11	46
	5 5	0.08 *	0.16 *	0.15	0.23	-1.71	0.09	-	-	1.69	0.10	46
	0 1	0.01 *	0.03 *	0.15	0.26	-1.86	0.06	-	-	1.77	0.08	46
	0 5	0.06 **	0.10 *	0.13	0.40	-1.82	0.07	-	-	2.01	0.05	46
Non-failure induced bank M&A												
CAAR	0 0	0.16% **	0.11% **	0.09	0.17	-1.99	0.05	2.13	0.03	1.95	0.05	142
	1 1	0.12%	0.10%	0.05	0.86	-0.90	0.37	1.06	0.29	1.02	0.31	142
	5 5	0.54% *	0.32% *	0.10	0.10	-1.69	0.09	1.63	0.10	1.83	0.07	142
	0 1	0.16%	0.17% *	0.06	0.61	-1.74	0.08	1.54	0.12	1.56	0.12	142
	0 5	0.43% **	0.32% *	0.10	0.12	-1.73	0.08	1.99	0.05	2.10	0.04	142
CAAB	0 0	0.00	0.01 *	0.12	0.03	-1.73	0.08	-	-	0.97	0.33	142
	1 1	0.01	0.02 *	0.12	0.04	-1.71	0.09	-	-	0.95	0.34	142
	5 5	0.05	0.07 *	0.11	0.05	-1.73	0.08	-	-	0.99	0.32	142
	0 1	0.01	0.01 *	0.12	0.04	-1.69	0.09	-	-	0.94	0.35	142
	0 5	0.03	0.04 *	0.11	0.06	-1.65	0.10	-	-	1.01	0.32	142

***, **, and * indicate statistical significance at the 1%, 5%, and 10% levels, respectively. *CAAR*: cumulative average abnormal return, *CAAB*: cumulative average abnormal beta shift, KS test: Kolmogorov-Smirnov test - p-value ≤ 0.05 indicates normal distribution at the 5% significance level, Wx-test: Wilcoxon signed rank test as non-parametric median test, Z-stat.: Boehmer test (parametric), t-test: parametric mean test.

The failures of IndyMac Bank and Washington Mutual are two events that are most widely known next to the very well-known failure of Lehman Brothers on 15 September, 2008. At the time of default IndyMac Bank had total assets of USD 32 Billion and deposits of USD 19 Billion, Lehman Brothers had total assets of USD 1.6 Trillion but no deposits due to its investment bank status, and Washington Mutual had total assets of USD 307 Billion and deposits of USD 188 Billion. Comparing these values to the average asset and deposit values of all other bank failures analyzed, their unprecedented sizes become evident: The average asset value of all other bank failures was around USD 995 Million and the average deposit value was around USD 810 Million.

We perform a separate analysis on these three bank failures to clarify whether they support our findings for the total sample of all bank failures dates. In contrast to the previous analyses, we use the available return series of up to 117 individual REITs, instead of REIT indices, to measure parallel risk and return shifts around the three failure dates. To control for the effect of regular bank M&A on a single event basis, we also perform an equivalent analysis on three bank M&As: The acquisition of Bank One by J.P. Morgan well ahead of the crisis announced January 14, 2004, the acquisition of ABN Amro by Bank of America just before the crisis announced April 23, 2007, and the acquisition of Wachovia by Wells Fargo announced at the peak of the crisis October 3, 2008.

We expect that our previous results should be supported. For the selected bank failures this implies positive return effects but no abnormal risk shifts relative to common equity behavior. For the control sample of regular bank M&As we also expect positive return effects with an absence of abnormal risk changes. Tables 4.7 and 4.8 summarize the main results on the combined risk and return analyses.

Altogether, the results of the analysis of the three single bank failures are comparable and rather supportive to our previous findings. The IndyMac Bank failure shows an impact on REITs that is generally in line with our previous findings. We find significantly positive *CAARs* at the event date and two further event windows, but also a highly significant positive abnormal risk shift is observable. The failures of Washington Mutual and Lehman Brothers also show highly significant positive *CAARs* around the failure dates with one exception of a negative *CAAR* at the Lehman failure date. Similar to the IndyMac Bank failure, they also incur highly significant positive beta risk shifts. Overall, the positive *CAARs* support our previous observations and the positive risk shifts indicate that especially large bank failures incur an abnormal increase in REIT systematic risk.

Table 4.7: Robustness Analysis on single REIT Behavior (I/II): Outstanding Bank Failures

Failure	Event date	[-x;+x]		Mean / Median *CAAR*			p-value		Mean / Median *CAAB*			p-value		n
IndyMac Bank	07.11.2008	0	0	1.65%	1.49%	***	0.00	z	0.02	0.02	***	0.00	w	115
		1	1	0.07%	0.68%		0.30	w	0.04	0.05	***	0.00	w	115
		5	5	1.33%	2.55%	***	0.01	w	0.12	0.14	***	0.00	t	115
		0	1	-0.38%	-0.51%		0.11	z	0.03	0.03	***	0.00	w	115
		0	5	1.30%	2.53%	***	0.01	w	0.09	0.10	***	0.00	w	115
Washington Mutual	09.25.2008	0	0	-0.20%	-0.82%		0.18	z	0.04	0.04	***	0.01	t	115
		1	1	0.37%	-0.57%		0.71	z	0.11	0.11	***	0.01	t	115
		5	5	5.79%	3.44%	***	0.00	z	0.39	0.39	***	0.01	t	115
		0	1	1.90%	1.01%	***	0.00	z	0.07	0.07	***	0.01	t	115
		0	5	3.77%	3.20%	***	0.00	z	0.18	0.21	*	0.07	t	115
Lehman Brothers	09.15.2008	0	0	-1.02%	-0.08%	**	0.03	z	0.02	0.02	***	0.00	t	115
		1	1	0.63%	2.23%	*	0.06	z	0.07	0.07	***	0.00	t	115
		5	5	3.29%	3.03%	***	0.00	z	0.38	0.35	***	0.00	t	115
		0	1	0.24%	1.90%		0.23	z	0.06	0.05	***	0.00	t	115
		0	5	2.59%	2.09%	***	0.00	w	0.28	0.22	***	0.00	t	115

***, **, and * indicate statistical significance at the 1%, 5%, and 10% levels, respectively. *CAAR*: cumulative average abnormal return, *CAAB*: cumulative average abnormal beta shift, Stat. p-value of either Boehmer test (z), t-test (t), or Wilcoxon test (w).

The control sample's results regarding the three large (regular) bank M&A dates shows strong ambiguity. Having addressed the critically different points in time of these M&A announcements with respect to the outbreak of the financial crisis this might serve as a starting point for a case-by-case analysis and interpretation of results. First, the pre-crisis takeover of Bank One by J.P. Morgan shows no significant abnormal return impact on REITs. This finding contradicts the results for the non-failure induced bank M&A robustness check which returned marginally positive abnormal returns at one to five percent significance levels. The fact that the respective M&A does not cause any abnormal risk shifts for REITs also contradicts previous evidence, as we found at least some positive abnormal beta shifts indicated by significant median values. Second, the acquisition of ABN Amro by Bank of America, announced just before the first real symptoms of the upcoming crisis appeared, shows an initially positive return effect at event date (1.07% at one percent significance level) which turns into a negative abnormal return in the symmetric [-5,+5] event window of -2.88% at the one percent significance level. Again, these results contradict the findings for the full sample of bank M&A between 1999 and 2010. And third, the impact from Wells Fargo announcing to take over Wachovia on REITs inverts the most recent observation in that at event date a negative *CAAR* of -2.10% at one percent significance level evolves into a strongly positive *CAAR* of 13.27% at the [-5,+5] event window. We also observe considerable positive risk shifts around regular bank M&A for the two latter cases. At least the risk observations support our previous findings from the robustness check on bank M&A. But, the former analysis' methodology allowed for a balancing of effects across 142 bank M&As implying that our results must not be interpreted as strongly contradictory.

Table 4.8: Robustness Analysis on single REIT Behavior (II/II): Outstanding Bank M&As

M&A	Event date	[-x;+x]	Mean / Median *CAAR*			p-value		Mean / Median *CAAB*			p-value		n
J.P. Morgan > Bank One	01.14. 04	0 0	-0.05%	-0.12%		0.91	z	0.00	0.00		0.85	w	89
		1 1	0.17%	-0.07%		0.74	z	0.00	0.00		0.70	w	89
		5 5	-0.26%	0.01%		0.77	w	-0.01	0.00		0.87	w	89
		0 1	0.13%	0.08%		0.48	z	0.00	0.00		0.65	w	89
		0 5	0.34%	-0.05%		0.42	w	-0.01	-0.03		0.61	w	89
Bank of A. > ABN	04.23. 07	0 0	0.93%	1.07%	***	0.00	w	0.03	0.03	***	0.00	t	113
		1 1	0.26%	0.14%	***	0.01	w	0.08	0.07	***	0.00	t	113
		5 5	-2.81%	-2.88%	***	0.00	w	0.28	0.23	***	0.00	t	113
		0 1	0.34%	0.48%	***	0.00	w	0.05	0.04	***	0.00	t	113
		0 5	-1.58%	-1.58%	***	0.00	w	0.14	0.11	***	0.00	t	113
W. Fargo > Wachovia	10.03. 08	0 0	-2.74%	-2.10%	***	0.00	w	0.03	0.03		0.23	t	117
		1 1	-2.71%	-2.03%	***	0.00	w	0.08	0.09		0.26	t	117
		5 5	9.22%	13.27%	***	0.00	w	0.42	0.41	**	0.05	t	117
		0 1	-2.68%	-2.09%	***	0.00	w	0.06	0.06		0.23	t	117
		0 5	5.01%	7.48%	***	0.00	w	0.29	0.25	*	0.03	t	117

***, **, and * indicate statistical significance at the 1%, 5%, and 10% levels, respectively. *CAAR*: cumulative average abnormal return, *CAAB*: cumulative average abnormal beta shift, Stat. p-value of either Boehmer test (z), t-test (t), or Wilcoxon test (w).

4.4.4. Multivariate Regressions of Focused REIT Indices on the All-REIT Index

Our final analysis consists of two sets on multivariate regressions: Dependent variables are the cumulative abnormal returns, *CARs*, and cumulative abnormal beta shifts, *CABs*, of the all REIT index on the exact event date. Independent variables are the respective *CARs* and *CABs* of all focused REIT indices. Our objective is to determine in how far the all REIT index based *CARs* and *CABs* are driven by the respective capital investment type REITs and the property type focused Equity REITs. In this analysis we concentrate on bank failures and respective shifts in risk-return profile of the recent financial crisis. As previous analysis have shown that failures during this period clearly drive total sample results and distinguish fully from those of the previous crisis we are convinced this is the best approach.

Due to space constraints we decide not to display all results but provide a condensed overview on the most relevant findings. As Table 4.9 shows we display the impact of all three capital investment type focused REITs on the all REIT index abnormal return and risk shifts. Then we select Industrial REITs and

Specialized REITs representatively for all property focused REITs because their coefficient values are the most extreme among all. Both multivariate regressions show that the Equity REIT index has the strongest impact on the FTSE NAREIT as coefficients are closely to the value of one at the one percent significance level and even the adjusted R^2 basically signals a value of one. The Mortgage REIT index also has strong and highly significant positive effect on the all REIT risk and return shift, however showing much lower valued coefficients. Interestingly, Diversified REITs again are not found between the Equity and Mortgage REITs but affect all REIT *CARs* negatively and do not impact all REIT *CABs*. Industrial and Specialized REITs, so as all other property focused REITs show a strong positive effect on the all REIT risk and return adjustments. These findings confirm previous evidence as Industrial REITs depart most in a former analysis and now show the lowest impact on the overall REIT index. Surprisingly, Specialized REITs even exceeds the impact of Residential, Retail, and Healthcare REITs on the all REIT index which we did not expect but in fact the differences are marginally as all coefficients are highly significant.

Table 4.9: Multivariate Regressions: Drivers of FTSE NAREIT CARs and CABs

Dependent variable: *CARs* [0] of FTSE NAREIT							*CABs* [0]					
Model		1	2	3	4	7		1	2	3	4	7
	α	0.00	0.00	0.00	0.00	0.00	α	0.00	0.00	0.01	0.00	0.00
		-0.3	*1.2*	*2.0*	*1.3*	*0.1*		*1.4*	*1.0*	*1.3*	*-0.6*	*0.3*
Independent variables: *CARs* [0] of focused REIT indices							*CABs* [0]					
Equity REIT	δ_1	1.00 ***					δ_{10}	0.96 ***				
		>9.9						*>9.9*				
Mortgage REIT	δ_2		0.61 ***				δ_{12}		0.40 ***			
			6.8						*7.6*			
Diversfied REIT	δ_3			-0.61 **			δ_{13}			0.08		
				-2.2						*0.7*		
Industrial REIT	δ_4				0.56 ***		δ_{14}				0.53 ***	
					6.2						*>9.9*	
Specialized REIT	δ_7					1.03 ***	δ_{17}					1.10 ***
						>9.9						*>9.9*
Nobs.	89						89					
R^2 (adj.)		1.00	0.48	0.18	0.72	0.89		1.00	0.53	0.00	0.76	0.88
F-stat. (p.-val.)		0.00	0.00	0.00	0.00	0.00		0.00	0.00	0.66	0.00	0.00
DW-stat.		1.98	2.20	2.29	1.60	2.06		0.45	0.74	0.76	0.66	0.36

***, **, and * indicate statistical significance at the 1%, 5%, and 10% levels, respectively; t-statistics in italic. *CAR*: cumulative abnormal return, *CAB*: cumulative abnormal beta shift. Models 5,6,8,9, and 10 concern multivariate regressions with Lodging, Office, Retail, Residential, and Healthcare REIT indices as independent variables. Results are not disclosed due to space limitations. Respective delta coefficients are all significant at the 1% level and their values are in-between those of Industrial and Specialized REITs. Detailed data is available upon request.

4.5. Conclusion

This research examines whether REITs perform differently than common equities when there are liquidity shocks induced by bank failures on monetary stability. The objective is to advance previous research by scrutinizing whether REITs deviate from common equity risk-return behavior when there are internal shocks to monetary stability that decrease primarily the supply of liquidity. This opposes previous evidence on REIT behavior to external monetary shocks such as monetary rate increases that primarily increase the price of liquidity.

Our null hypothesis states that REITs' risk-return profile should not behave significantly different from other equities as in the case of unexpected monetary rate changes. Our alternative hypothesis bases on the safe haven theory of beneficial real estate valuation during times of crisis and presumes a positive REIT valuation around bank failures. Our empirical results for the total sample show that REITs generally experience positive abnormal returns (averaging 0.93%) around bank failure dates while systematic risk levels stay relatively stable. This result supports the safe haven theory by Mueller and Mueller (2003) and indicates that REITs presumably react more like real estate than common equities during crisis.

Next we analyze the unique drivers of these results and extend our analysis. The first subsample controls for two time periods: the *New Economy: boom bust of 1999-2004* versus the *Real Estate Securitization: boom bust of 2005-2010.* The first period results contradict the overall results with some negative abnormal returns and some negative abnormal risk shifts but still rather similar behavior to common equities. The second period (which has the majority of failures and covers the great financial crisis) strongly supports and drives the overall results that REITs outperform common equities during banking crisis.

Furthermore we find that Equity REITs in contrast to Mortgage REITs and Diversified REITs show strong positive abnormal returns around bank failures. This provides evidence that investors value direct equity investments in real estate during times of high market uncertainty. Regarding property focused Equity REITs we can confirm expectations that the low risk REITs concentrating on retail, residential, and healthcare property investments outperformed their peers. These findings clearly provide evidence for the save haven theory of real estate investments as hard assets in times of high market uncertainty.

Finally, the twofold robustness analysis approach on the total sample finds a marginally positive return effect during *regular bank M&A* events. Non-failure Fridays though do not show a considerable bias on our results. The subsequent analysis on single major bank failures of the year 2008 supports our previous findings for total sample and the *Real Estate Securitization*-phase providing further evidence for positive abnormal return effects. The single bank M&A events return a variety of REIT risk-return profile shifts which neither support nor clearly contradict our previous results. In sum, the variety of analyses and results supports our methodological approach to first perform total sample analysis across two different economic cycles so that single event effects are balanced out.

Concluding, we find internal monetary system shocks, like bank failures, seem to have a different effect on REITs versus the general stock market, compared to external shocks like monetary rate increases, even though, both shocks decrease market liquidity. In our opinion, these findings support the *safe haven*-theory of real estate valuation in turbulent times. While monetary rate increases are commonly triggered during times of strong growth, implying high market confidence regarding return prospects, bank failures occur unexpected and during economic crises causing market uncertainty about the near future forcing stock price declines.

We also evaluate bank bailouts and REIT returns of high interest to scholars of financial and real estate economics. This should represent value added research as it could shed light on the other side of the recent financial crisis. An interpretation of bank bailouts as a positive shock to market liquidity could enable a view on the probable asymmetry of positive versus negative shocks to monetary markets and REIT returns.

The three main limitations of this study are: First, we only provide evidence for the U.S. market; second, our analyses is limited to the two most recent crises which might cause deviations in implications compared to previous crises; and third, the current accumulation phase of U.S. bank failures was not over as of 2010 when this data was analyzed. Future research could extend this study with additional time periods and variables to advance insights into the cause-effect relationship between bank failures and REIT returns.

5. Bank Bailouts: REITs and their Performance as Financial Stock[12]

5.1. Introduction

In the turbulent times of the still ongoing financial crisis, starting in 2007, the financial sector, first and foremost banks, and the real estate sector, which is commonly believed to have triggered the global crisis through the subprime sector, have been in the center of the public interest. The banking crisis and the fear of a fatal breakdown of the economic system have caused the government to launch, widely accompanied by a wide public discussion, the most extended bailout program in the history of the United States of America, creating the USD 700 billion Troubled Asset Relief Program (TARP) to purchase failing bank assets. The major goals of this program were the stabilization of the economy, the improvement of the markets' liquidity and to reconstitute investors' confidence (Paulson, 2008).

Although REITs are commonly classified as financial stocks along with banks, many consider them to be an asset class of its own, due to their special characteristics and differences from common equities. REITs are highly dependent on external financing, which might on the one hand hinder their performance during times of market uncertainty, but on the other hand might drive their performance following a recapitalization of the banking sector.

There has been a brought discussion on the moral hazard involved with government bailout programs, and the danger of encouraging excessive risk taking behavior. Looking at REITs, previous research has shown that REITs' risk-return behavior appears to be a combination of real estate and common equity market behavior, which does not allow for a definite allocation to any group. All in all, we are convinced that a detailed analysis of REITs' risk-return behavior compared with other financial stocks and relative to common equities in the event of bank bailouts represents worthwhile research effort and adds well to current literature.

---

[12] I thank Jan-Willem Olliges for his major contribution to this study concerning the industry specific literature review, data analysis, and the study structure. I am also grateful to Dirk Schiereck and Glenn Mueller for valuable comments and insights. Special thanks go to the conference participants at the annual meeting of the American Real Estate Society 2011 in Seattle for valuable feedback.

Regarding behavioral expectations of REITs' risk-return movements around bank bailouts, we would expect a behavior similar to common equities but find negative valuation effects. The recapitalization of the banking sector, resulting in restored liquidity of the market should favor relatively risky investments such as equities, and hereby including REITs, compared to low risk investments such as government bonds and commodities, e.g., precious metals. Because REITs are strongly dependent on external financing they might especially profit from restored market liquidity. This valuation effect on REITs constitutes our null hypothesis.

The alternative hypothesis implies a relatively negative valuation effect for REITs. As investors commonly interpret REITs as a safe haven during financial crisis, REITs might be outperformed by other equities in the aftermath of a market recapitalization. Moral hazard symptoms, encouraging risk taking, might favor assets that show less *hard asset* characteristics than REITs and appear to be more risky and hence more promising. Consequently, the alternative hypothesis assumes that REITs do not follow common equities behavior in the case of monetary shocks through bank bailouts as they do in the case of monetary shocks through unexpected monetary rate changes.

The effects of monetary shocks to the market have extensively been researched. Bernanke and Kuttner (2005) researched the impact of a change in monetary policy and found that unexpected decreases in the monetary rate let to an increase in stock prices. In their research Bredin, O'Reilly, and Stevenson (2007) focus on the effects these monetary rate adjustments might have on REITs. They also state that REITs, due to their special characteristics, are a good estimator for common equities' behavior. Diamond and Rajan (2005) showed that, when banks fail, this can potentially be contagious and a consecutive shrinking of the market's liquidity pool might trigger liquidity shortages, representing a monetary shock to the market.

We are curious to investigate whether REITs show comparable results to common equities in the case of bank bailouts and if their behavior depends critically on the type of monetary shock to the market. Monetary rate increases, as well as bank failures, result in a limitation of the market's liquidity pool. Nonetheless, they generally occur in completely opposing market situations. Monetary rates are generally increased in times of positive market prospects, whereas bank failures typically occur at the peak and in the aftermath of an economic crisis. Our earlier research documents that REITs show a positive valuation effect in relation to the market in the case of bank failures and hence react differently compared to the case of a monetary rate increase. We are curious to find out whether, in the case of bank bailouts, REITs show comparable behavior to other equities. This expectation is based on the fact that generally the market is in the same state when bank bail-outs occur or monetary rates are decreased. However, it is also a valid assumption to expect negative valuation effects, as bank bailouts represent the opposite outcome for banking crises compared to bank failures, which caused positive REIT valuation effects.

We analyze the impact of bank bailouts on REITs as financial stock in the U.S. with respect to any shifts in prevalent risk-return profiles. Consistent with a previous study on bank failures the research period extends from January 1999 until July 2010 and covers US-REITs. In addition, a representative sample of other financial stocks is analyzed for robustness. Although the analyzed time period reaches back to 1999, the actual observations were made from January 2008 until July 2010, due to the fact that no bank bailouts occurred in the time period from January 1999 until January 2008. Within the observation period 700 banks where bailed out on 49 different dates. Eliminating dates that overlap with bank failure dates, seven bank bailout dates are left. In order to obtain a number of events that can enable statistical significance of results, the research is carried out on single stock level instead of on index level. Hence, the analysis is conducted for 121 REITs and for 229 other financial stocks. Diverse subsamples where set up in order to investigate whether different types of REITs or other financial stocks are affected in a different manner.

For all events we analyze short window abnormal returns and abnormal risk shifts to verify whether bank bailouts have affected REITs' risk-return profile differently relative to common equity markets and other financial stocks, and whether we can observe different reactions compared to the findings on bank failures in previous research.

The paper is structured as follows: In part 5.2 we review the literature; in part 5.3 we explain our data selection and the empirical methodology; in part 5.4 we present our results; part 5.5 concludes.

5.2. Literature Review

5.2.1. Market Liquidity, Bank Bailouts, and REIT Behavior

Bernanke and Kuttner (2005) investigate the effect of changes in monetary policy as a monetary shock on equity prices and find that an unanticipated cut of the federal fund rate leads to an increase in stock prices. They furthermore find that the largest part of response in stock prices is accountable to the revision of cash flows but only a very small portion is attributable to the changes in expected real interest rates. In their research Bredin, O'Reilly, and Stevenson (2007) focus on the effects that monetary rate changes may have on REIT returns and volatility.

Their results are in line with previous evidence on broader equity markets and show, that REITs show a strong response to unanticipated changes in monetary policy in both, first and second moments of REIT returns. Their findings support the expected trend of the effect, hence a positive return and a negative volatility reaction for a cut in the monetary rate.

The assumption, that banking crises can also be seen as a monetary shock is supported by the findings of Diamond and Rajan (2005). They state, that banking crises, especially when resulting in bank failures, can potentially shrink the market's liquidity pool, causing liquidity shortages. Furthermore they state, that bank failures can be contagious which may eventually lead to a collapse of the financial system. Gorton and Huang (2004) state in their research on market liquidity, efficiency, and bailout policies, that market liquidity can effectively be strengthened or restored through bailing out distressed banks and the government can actually improve welfare by doing so. Concluding, it can be said that banking crises as well as liquidity injections by the government can be seen as monetary shocks comparable to changes in the federal fund rate.

REITs rely heavily on external financing. Dell'Ariccia, Detragiache, and Rajan (2008) found that banking crises can externally hinder real activity, especially in those sectors that are more dependent on external financing. This might lead to the expectation that REITs profit from the positive signaling effect of a bank bailout, strengthening the market's liquidity pool. Furthermore, Giannetti and Simonov (2009) found that the recapitalization of a bank leads to positive abnormal returns for the bank's customer companies. It remains the question whether bank bailouts as a positive market signal provoke the same effect on broader equity markets or even favor capital intensive industries like REITs. In their investigation of stock market responses on bank restructuring policies Klingebiel et al. (2001) discover that bank stock prices as well as non-financial stocks respond positively to the announcements of government guarantees for bank liabilities. However, non-financial companies react negatively when public recapitalization schemes aim at strengthening banks' capital base and generous liquidity support programs are announced. Bank stocks do not show a similar reaction. Klingebiel et al. (2001) conclude from these results, that "announcements of deposit guarantees may provide comfort to market participants during a crisis of confidence, but [...] the provision of public funds for bank bailouts were not necessarily viewed as credible means to restore the health of the financial sector."

Bank bailouts are often accompanied by a public discussion about the moral hazard this might provoke. There is an ongoing discussion whether banks should generally be bailed out and if so, what would be the optimal way, concerning the social cost - moral hazard - trade-off (see Goodhart and Huang, 1999). Standard

LOLR (lender of last resort-function of the central bank) practice assumes that ex-ante commitment to assist troubled institutions provokes excessive risk taking. This moral hazard should therefore be avoided (see Cordella and Yeyati, 2003). Furthermore Hakenes and Schnabel (2010) find that the bailout of banks leads to higher risk taking among the protected bank's competitors, independently of market transparency. Nonetheless, Cordella and Yeyati (2003) also state, that the right bailout design might, under certain circumstances, also ex-ante, lower the banks' appetite for risk. Concluding, this poses the question what effects bank bailouts might have on first and second moments of return of REITs and other financial stocks compared to the general equity market.

When investigating what generally influences REIT returns, Ghosh, Guttery, and Sirmans (1998) on the one hand found that REIT prices do respond to real estate related news, whereas Giliberto (1990) on the other hand found that the correlation between non-securitized real estate returns and REIT returns was close to zero. Ghosh, Miles, and Sirmans (1996) state that REITs are strongly influenced by stock market volatility in the short-term but their long-term return is derived from direct real estate growth. However, a constraint is the fact that the interdependencies are never constant. This is supported by Clayton and MacKinnon (2001) who find that the link between REITs, real estate, and other asset classes constantly changes.

Since REITs have recently been added to three S&P general market indices they co-move more with common equities than their real estate fundaments justify, as has been discovered by Ambrose, Lee, and Peek (2007) in their analysis of spillover effects across asset classes. Finally, Basse and Friedrich (2009) reexamine the relationship between REITs and utility stocks in the current financial crisis and conclude that the crisis led to a structural change in this relationship. Their findings show that investments in REITs have become much more risky in times of crisis.

5.2.2. Contribution to the Literature and Hypotheses

The current banking crises and the discussion about the possible outcomes, on the one hand bank failures and on the other hand the government bailing out distressed institutions, as it has been the case under the USD 700 billion Troubled Asset Relief Program, has shown that market liquidity and its implications on real market behavior is of high interest to the public as well as to scholars of both, financial economics and real estate economics. As effects of federal fund rates changes, as an external monetary shock, have been broadly researched, we are now keen to examine the effects of banking crises as an internal monetary shock to the system. Research on monetary rate adjustments has shown that REITs show comparable risk-return behavior in size and direction to common equities. As previous research has also elucidated, banking crises can also be seen as an internal type of monetary shock, which although might potentially show differing results. As our earlier research on bank failures has shown, this indeed might be the case and we are now eager to investigate the effects bank bailouts trigger for REITs and other financial stocks. We believe analyzing bank bailout events as a shock to monetary stability and its unique impact on REITs and other financial stocks adds well to the literature.

Regarding behavioral expectations of REITs' return movements around bank bailouts, in line with the research by Bredin, O'Reilly, and Stevenson (2007), it is reasonable to expect behavior similar to common equities, which constitutes our null hypothesis. The occurrence of return behavior comparable to that of common equities would seem to be reasonable as bank bailouts may lead to a comparable increase in market liquidity as caused by monetary rate decreases. This argument finds further support by the fact that both monetary shocks generally occur when the market is in the same state of an economic cycle, in this case a market downturn.

The alternative hypothesis, this is REITs showing different return behavior compared to common equities, builds on previous research on REIT behavior in the event of bank failures. Here, positive abnormal returns of REITs relative to common equities are observed. In the case of bank bailouts, as the counterpart of bank failures, this could lead to corresponding results. Henceforth, our alternative hypothesis constitutes that REITs as hard assets now underperform common equities, showing negative abnormal returns.

5.3. Data and Methodology

5.3.1. Data Selection and Structure of the Analysis

The sample of the U.S. bank bailout announcements is taken from the list of bailed out banks by the U.S. Department of the Treasury (2011). Consistent with our previous research on bank failures, we first consider all U.S. bank bailout dates between January 1999 and July 2010. Nonetheless, bank bailouts only occurred after January 2008 leaving us with a relatively short observation period which might potentially bias our results. The total sample of bank bailouts is composed of 49 event dates reflecting a total of 700 single bank bailouts totaling about USD 204 billion of invested capital by the U.S. Department of the Treasury.

Cross checking with bank failure dates to eliminate overlapping, we find that only seven bank bailout dates are left. To overcome this problem, that with only seven event dates it is hardly possible to obtain statistically meaningful results when analyzing REIT indices, we conduct our analysis on single REIT level for 121 REITs as well as on single company level for other financial stocks. In total 229 other financial institutions are analyzed. The S&P 500 Composite Index is used as an overall market return index. The stock return data is retrieved from Datastream. Respective return series are used from January 1st, 2007 until July 31st, 2010.

We apply standard event study methodology to measure short term abnormal returns around bank bailout dates. To obtain comparable results on a risk perspective we use a beta-aligned event study approach. Determining daily REIT beta measures by applying a simple GARCH (1,1) model enables us to capture market adjusted abnormal beta shifts within short-term windows, equivalent to the standard event study.

The 121 REITs that are investigated in this analysis divide into REITs from eight different subsectors which are composed as follows: 10 industrial, 11 lodging, 12 office, 22 retail, 14 residential, 12 healthcare, 10 specialized, and 16 diversified REITs. Furthermore, the REITs can be divided into 91 Equity REITs, 14 Mortgage REITs, and the 16 Diversified REITs. The Diversified REITs are not accounted for in the Equity REIT subsample since they may contain both, equity and mortgage real estate investments. In the course of the investigation also a representative sample of 229 other financial stocks in analyzed. This sample is composed of 30 asset management companies, 119 banks, 19 brokers and 61 insurances.

In order to gain a more precise insight on REIT and other financial stock behavior we perform return and risk event studies on every individual subsample. For an event window [-1, +1 day] we find the following *CAARs* (mean values): -3.34% for Lodging REITs, -2.51% for Office REITs, -1.61% for Retail REITs, -3.12% for Residential REITs, -1.94% for Healthcare REITs and -1.30% for Industrial REITs. For Specialized REITs and Diversified REITs we do not find statistically significant results. So, the overall result of a negative return effect is supported almost throughout every subsample, which is in line with the observation of a negative return effect for Equity REITs of -2.08%. Mortgage REITs do not show any statistically significant return effect.

When looking at the risk effect, the picture is not quite as clear as with the return effect. For the total sample we do not find any statistically significant abnormal shift in risk regarding the means (the medians nonetheless show a positive risk shift). For Lodging and Healthcare REITs we observe a negative risk shift, Office REITs, Residential REITs, Specialized REITs, and Diversified REITs show a positive risk shift.

When scrutinizing for a possible "Financial Stock" biasing effect we find that the negative return effect is observable in numerous event windows (total sample, as well as subsamples), although the picture is not as clear as it is for REITs. When looking at the risk effect we observe a highly significant positive risk shift throughout basically every event window in every subsample.

Comparing these results with our findings from previous research on bank failures, where we observed a positive return effect and no significant risk shift, our findings add well to the literature.

Five observations might uncover as potential biases for our analyses:

First, at each specific bank bailout date up to 49 banks (on average 27 banks) were bailed out by the U.S. government at the same time. This is the reason why we denote the events as bank bailout dates instead of bank bailouts.

Banking crises have two possible outcomes: first, being the recapitalization by the government, namely the bank bailout and the second, being the bank failure which commonly results in bank M&A because former competitive financial institutions continue the failed banks' operations and accounts. Hence, we use the results of our previous research on bank failures as a robustness check for this analysis.

Due to the prevalent accumulation of bank bailouts in short time periods the average time gap between the analyzed bank bailout dates is 18.6 days. This could lead to a certain overlap of effects in relatively short event windows. We control for this biasing issue by narrowing the maximum event window size to a [-2,+2] days around the specific event date.

Although we take into consideration the time period from January 1999 until July 2010, our observations account for a relatively small period of time, namely January 2008 until July 2010, since bank bailouts only occurred within this period of time.

Eliminating all bank bailout dates that overlap with bank failure dates only seven out of 49 dates remain which leaves us with a very small sample. We overcome this biasing issue by testing these seven event dates on single REIT level, respectively single financial stock level, to provide a sufficient amount of data points.

5.3.2. Methodology of the Return and Risk Event Study

In our analysis on REITs and other financial stocks we investigate both, first and second moments of return. Therefore, our approach must be twofold. First, we look at the return component and secondly, we investigate the risk shift.

For the return part we apply a standard event study methodology similar to MacKinlay (1997). Hereby, for the regression, we use the market model approach according to the Capital Asset Pricing Model (CAPM) by Sharpe (1964) and Lintner (1965). In order to obtain case specific betas *(β)* as systematic risk estimators, we perform a simple Ordinary Least Squares (OLS) regression 30 days prior to each event over a 250 trading days estimation period. As a product of these betas and the actual market returns we are now able to forecast expected REIT returns throughout the event windows. Subtracting the expected returns from the actually observed returns, we obtain the Abnormal Returns *(AR)*. For every single event the abnormal returns are cumulated accordingly to the varying event window size which gives us the Cumulative Abnormal Returns *(CAR)*. The Cumulative Average Abnormal Returns *(CAAR)* are derived by computing the arithmetic and geometric averages of the *CARs*, accounting for the specific number of events. We estimate the *CAARs* for the event windows [0], [-1,+1], [-2,+2], [-1,0], [-2,0], [0,+1] and [0,+2]. Since the average time span between two events is only 18.6 trading days and the smallest gap only 5 days, longer event windows are not accounted for, in order to minimize biasing effects through overlap of events and effects.

A Kolmogorov-Smirnov test is applied to verify whether the *CARs* are normally distributed (see Massey, 1951). The *CAAR* means are tested for difference from zero with a parametric Boehmer test (see Boehmer, Musumeci, and Poulsen, 1991). For the medians this is done by applying the Wilcoxon signed rank test.

For the risk analysis we also use an event study methodology that enables us to determine abnormal risk shifts and test them statistically for difference from zero. The mathematical approach to this is very similar to the standard event study applied for the return analysis. Our risk event study model consists of five steps. In the first step, daily REIT and other financial stock betas have to be calculated. In the next step, expected betas have to be estimated around bank bailout dates from before the respective events. The third step consists of subtracting the observed betas from the expected betas in order to obtain Abnormal Beta shifts *(AB)*. Next, the Cumulative Abnormal Beta shifts *(CAB)* and the Cumulative Average Abnormal Betas shifts *(CAAB)* are calculated. In the fifth and last step, the *CAABs* are tested for difference from zero.

Measuring time-variant betas contradicts one of the basic assumptions of the Capital Asset pricing model *(CAPM)*, in which it is assumed that beta is stable over time. Nonetheless, many financial economists like Fabozzi and Francis (1978), Sunder (1980), Bos and Newbold (1984), Collins, Ledolter, and Rayburn (1987) and Jagannathan and Wang (1996) have argued against the assumption of beta stability and shown that beta is actually stochastic and thus conditional.

We apply the generalized autoregressive conditional heteroscedasticity (GARCH) approach in order to determine time-varying volatilities and beta measures. The GARCH model is one of the most commonly applied methods to observe changes in risk levels. The technique was developed by Engle (1982) and later on refined by Bollerslev (1986). It was also Bollerslev who advanced the model to the Multivariate GARCH (M-GARCH) (see Bollerslev, 1990). Giannopoulos (1995) shows that GARCH models can not only be used to measure total risk but also to divide into the two components, systematic and unsystematic risk. Brooks, Faff, and McKenzie (1998) investigate the forecasting accuracy of three techniques regarding time-variant beta: a M-GARCH, a time-varying beta market model suggested by Schwert and Seguin (1990), and the Kalman filter for a sample of returns on Australian industry portfolios. Furthermore, Faff, Hillier, and Hillier (2000) applied a M-GARCH approach to model time-varying beta in their study discussing alternative beta risk modeling techniques. They define an additional assumption of a constant correlation coefficient between the returns of an asset and respective market returns. The behavior of this correlation coefficient is core to our risk event study determining *CAABs*.

The simple GARCH (1,1) model determines a specific asset's time-varying variance as a weighted sum of a long-term average variance rate, Φ_i, the previous day's squared return, $u_{i,t-1}^2$, and the previous day's variance, $\sigma_{i,t-1}^2$. The weighting factors γ_1, γ_2, and γ_3 are determined by applying the iterative Maximum Likelihood technique as discussed by Hull (2006).

$$\sigma_{it}^2 = \gamma_1 \Phi_i + \gamma_2 u_{i,t-1}^2 + \gamma_3 \sigma_{i,t-1}^2 \tag{5.1}$$

Obtaining beta measures as next step we apply the M-GARCH model that observes the actual betas as a product of a constant correlation coefficient, ρ_{im}, between the specific stock (REIT and other financials) returns and the market returns (S&P 500 Composite) and the time-varying stock standard deviation, σ_{it}, divided by the time-varying market standard deviation, σ_{mt}.

$$\beta_{it} = \frac{\rho_{im}\sigma_{it}}{\sigma_{mt}} \tag{5.2}$$

Finishing the empirical groundwork, we have the following data: Daily volatility measures for all single REITs, all single other financial stocks and the S&P 500 Composite; daily correlation coefficients for all single REITs with the S&P 500 Composite and all other financial stocks with the S&P 500 Composite, based on a one year lag structure; and daily beta measures for all single REITs and all single other financial stocks in relation to the S&P 500 Composite.

Equivalent to beta for the standard return event study we utilize the correlation coefficient 30 days prior to the event for our systematic risk estimates, $E(\beta_{it})$. REIT and market volatility measures remain time variant as the market return does in the standard event study. The abnormal beta shift, *AB*, is then determined as the difference between expected and measured beta from up to two days before to up to two days after the event.

$$AB_{it} = E(\beta_{it}) - \beta_{it} = \frac{\rho_{im}\sigma_{it}}{\sigma_{mt}} - \beta_{it} \tag{5.3}$$

Cumulating *ABs* along the differently sized event windows [0], [-1,+1], [-2,+2], [-1,0], [-2,0], [0,+1], and [0,+2] we obtain cumulative abnormal beta shifts, *CAB*. The arithmetic mean of *CABs* across the number of events, *n*, per analysis returns the cumulative average abnormal beta shifts, *CAAB*. Again, the Kolmogorov-Smirnov test is used to determine if these *CABs* are normally distributed. The non-parametric Wilcoxon signed rank test and the parametric t-test are applied to test the *CAABs* for statistical difference to zero.

5.4. Results

5.4.1. Bank Bailouts' Effect on REITs and Other Financial Stocks

We begin our analysis by measuring the effect of bank bailouts on risk-return profiles of our total sample of REITs. Before that we will derive expectations regarding the reaction of REITs in the event of bank bailouts. Expectations will also be formulated before comparing the results with the total sample of other financial stocks, as well as before looking into the different subsamples.

Our first analysis considers the entire sample of 121 REITs and observes the effects bank bailouts may have on the REITs' risk-return profile. We expect bank bailouts as a monetary shock to have a similar effect on REITs as the shock through federal fund rate adjustments has. This means, that we expect REITs to react congruent to common equities and therefore that no significant differences in the risk-return profile can be observed compared to the market which is represented by the S&P 500 Composite index. However, as Table 5.1 shows significant negative abnormal returns (between -0.25% and -1.35%) are found for six of the seven event windows. The risk profile however, remains unchanged when looking at the means. The medians show a significant risk increase, the Kolmogorov-Smirnov test however indicates, that in this case the means should be looked at, which show no statistically significant difference from zero. Accordingly, the total sample analysis indicates that bank bailouts induce a negative return shift for REITs compared to common equities. The REITs' systematic risk appears to show no difference from common equities behavior. This results in a negative valuation effect for U.S. American REITs in the event of bank bailouts in comparison to the common equity markets. Thus, our alternative hypothesis is supported at an overall level.

Table 5.1: Risk/Return Event Study: Total Sample of REITs

	Abnormal returns, *CAAR*												
	[-x; +y]	Mean		Median		KS-test	p	Wx-test	p	t-test[1]	p	N
Total sample REITs	0 0	0,36%		-0,12%		0,10	0,00	-0,65	0,52	1,40	0,16	847
	1 1	-1,35%	***	-1,84%	***	0,09	0,00	-5,76	0,00	-4,76	0,00	847
	2 2	-1,32%	***	-1,89%	***	0,09	0,00	-4,84	0,00	-4,19	0,00	847
	0 1	-0,62%	**	-0,55%	***	0,09	0,00	-3,27	0,00	-2,27	0,02	847
	0 2	-0,71%	***	-0,97%	***	0,09	0,00	-3,82	0,00	-2,76	0,01	847
	1 0	-0,37%	***	-0,77%	***	0,10	0,00	-2,88	0,00	-2,47	0,01	847
	2 0	-0,25%	*	-0,77%	**	0,10	0,00	-2,22	0,03	-1,77	0,08	847
	Abnormal beta shifts, *CAAB*											
	[-x; +y]	Mean		Median		KS-test	p	Wx-test	p	t-test[1]	p	N
	0 0	-0,01		0,02	***	0,26	0,00	-6,57	0,00	-1,27	0,20	847
	1 1	-0,04		0,06	***	0,26	0,00	-6,30	0,00	-1,35	0,18	847
	2 2	-0,06		0,11	***	0,26	0,00	-6,21	0,00	-1,35	0,18	847
	0 1	-0,03		0,04	***	0,26	0,00	-6,33	0,00	-1,41	0,16	847
	0 2	-0,04		0,06	***	0,26	0,00	-6,25	0,00	-1,43	0,15	847
	1 0	-0,02		0,04	***	0,26	0,00	-6,38	0,00	-1,25	0,21	847
	2 0	-0,04		0,06	***	0,26	0,00	-6,27	0,00	-1,25	0,21	847

***, **, and * indicate statistical significance at the 1%, 5%, and 10% levels, respectively. *CAAR*: cumulative average abnormal return, *CAAB*: cumulative average abnormal beta shift, KS test: Kolmogorov-Smirnov test - p-value ≤ 0.05 indicates normal distribution at the 5% significance level, Wx-test: Wilcoxon signed rank test as non-parametric median test, [1]t-test: Parametric mean tests: Boehmer test for *CAARs*; Student's t-test for *CAABs*.

These results add well to the findings of our earlier research on bank failures. The respective analysis showed generally positive valuation effect for U.S. American REITs in the event of bank failures while risk behavior showed no deviation from common equities.

REITs are generally considered to be financial stock. In order to broaden our insights on the behavior of REITs in the case of monetary shocks, we also investigate the behavior of other financial stocks in the case of bank bailouts. Again our expectation is that financial stocks show a highly similar behavior to common equities, which again makes us expect not to observe any abnormal returns or abnormal beta shifts. As Table 5.2 shows, against our expectation, we find negative return effects for four of the seven observed event windows (between -0.30% and -0.79%). For financial services institutions (FSIs) we observe an abnormal increase in risk relative to the market for all event windows. Regarding the return analysis REITs and FSIs show a somewhat comparable behavior. The risk analysis although shows a clearly diverging behavior.

Table 5.2: Risk/Return Event Study: Total Sample of FSIs

Total Sample FSIs											
Abnormal returns, *CAAR*											
[-x; +y]	Mean		Median		KS-test	p	Wx-test	p	t-test[1]	p	N
0 0	-0,47%	***	-0,09%	***	0,12	0,00	-2,85	0,00	-4,35	0,00	1603
1 1	-0,01%		-0,07%		0,10	0,00	-1,26	0,21	-1,05	0,29	1603
2 2	-0,30%	**	-0,15%	**	0,10	0,00	-1,95	0,05	-2,17	0,03	1603
0 1	-0,72%	***	-0,33%	***	0,11	0,00	-5,13	0,00	-5,29	0,00	1603
0 2	-0,79%	***	-0,52%	***	0,10	0,00	-4,99	0,00	-5,05	0,00	1603
1 0	0,24%		0,36%		0,10	0,00	-1,62	0,11	0,18	0,86	1603
2 0	0,02%		0,02%		0,09	0,00	-0,55	0,58	-0,60	0,55	1603
Abnormal beta shifts, *CAAB*											
[-x; +y]	Mean		Median		KS-test	p	Wx-test	p	t-test[1]	p	N
0 0	0,03	***	0,03	***	0,11	0,00	-15,33	0,00	11,63	0,00	1603
1 1	0,10	***	0,10	***	0,11	0,00	-15,28	0,00	11,76	0,00	1603
2 2	0,16	***	0,16	***	0,11	0,00	-15,30	0,00	11,93	0,00	1603
0 1	0,06	***	0,06	***	0,11	0,00	-15,20	0,00	11,51	0,00	1603
0 2	0,10	***	0,10	***	0,11	0,00	-15,30	0,00	11,61	0,00	1603
1 0	0,07	***	0,06	***	0,11	0,00	-15,35	0,00	11,90	0,00	1603
2 0	0,10	***	0,09	***	0,11	0,00	-15,29	0,00	12,04	0,00	1603

***, **, and * indicate statistical significance at the 1%, 5%, and 10% levels, respectively. *CAAR*: cumulative average abnormal return, *CAAB*: cumulative average abnormal beta shift, KS test: Kolmogorov-Smirnov test - p-value ≤ 0.05 indicates normal distribution at the 5% significance level, Wx-test: Wilcoxon signed rank test as non-parametric median test, [1]t-test: Parametric mean tests: Boehmer test for *CAARs*; Student's t-test for *CAABs*.

In order to verify whether the results for the return part, that at first sight showed a comparable reaction, are truly alike, we conduct a mean-difference test. As Table 5.3 shows, we find strong evidence that the bank bailouts affect the return behavior of REITs and other financial stocks differently. The mean return reactions of REITs and other financial stock significantly differ in four of the seven event windows – for medians this is the case for five out of the seven event windows. However, there is no clear trend observable whether REITs or other financial stock are affected more negatively concerning the return behavior.

Table 5.3: Mean Difference Test on CAAR of REITs and FSIs

[-x; +y]	*CAAR* REITs		*CAAR* FSIs		Diff. in Returns	Wx-test	p-value		t-test	p-value	
0 0	0,36%		-0,47%	***	-0,84%	2,02	0,04	**	-3,28	0,00	***
1 1	-1,35%	***	-0,01%		1,34%	-4,76	0,00	***	3,44	0,00	***
2 2	-1,32%	***	-0,30%	**	1,02%	-3,68	0,00	***	2,16	0,03	**
0 1	-0,62%	**	-0,72%	***	-0,10%	-0,20	0,84		-0,32	0,75	
0 2	-0,71%	***	-0,79%	***	-0,08%	-1,06	0,29		-0,22	0,83	
1 0	-0,37%	***	0,24%		0,60%	-3,36	0,02	**	1,78	0,08	*
2 0	-0,25%	*	0,02%		0,27%	-1,89	0,00	***	0,68	0,49	

***, **, and * indicate statistical significance at the 1%, 5%, and 10% levels, respectively. *CAAR*: cumulative average abnormal return. Wx-test: (Wilcoxon rank sum test) non-parametric median test, t-test: (Student's t-test) parametric mean test.

5.4.2. Bank Bailouts' Effect on Focused REITs

With the analysis of various subsamples we target to find determinants that can add explanatory value to the results observed in the total sample analysis. We perform risk-return event studies for eight different REIT investment property types: Industrial, Office, Retail, Residential, Healthcare, Lodging, Specialized, and Diversified REITs. Furthermore, we compare the results of Equity REITs (consisting of Industrial, Office, Retail, Residential, Healthcare, Lodging, and Specialized REITs) and Mortgage REITs.

We start by comparing Mortgage REITs to Equity REITs. Equity REITs invest directly into own stakes in properties, whereas Mortgage REITs invest into loans backed by real estate. Since the total sample is mainly composed of Equity REITs and this sample shows negative return effects while risk levels are stable, we do now expect to make similar observations for Equity REITs. Mortgage REITs, due to their investment into loans, and hence their proximity to the lending sector should stay unaffected in relation to the market or might even profit from bank bailouts. See Payne and Mohammadi (2004) for a previous study on REIT capital investment types and cointegration with equity and bond markets that shows that shocks may indeed disseminate quickly across REIT markets.

For all subsample analyses we have chosen to display the results for the exact event day, the [-1,+1] and the [-2,+2] event window because they display the most significant results in a condensed form. Reporting all event windows would exceed the limits of this paper. Important to mention, we do not leave out any contradictory results.

Table 5.4: Risk/Return Event Study: REIT-type Subsamples (I/II)

REIT Type		[-x; +y]	Mean		Median		KS-test	p	Wx-test	p	t-test[1]	p	n
Equity	*CAAR*	0 0	0,46%		-0,07%		0,10	0,00	-0,89	0,37	1,59	0,11	637
		1 1	-2,08%	***	-2,19%	***	0,07	0,01	-6,67	0,00	-6,17	0,00	637
		2 2	-2,36%	***	-2,39%	***	0,09	0,00	-6,33	0,00	-5,76	0,00	637
	CAAB	0 0	-0,01		0,02	***	0,25	0,00	-5,05	0,00	-1,04	0,30	637
		1 1	-0,03		0,05	***	0,25	0,00	-4,82	0,00	-1,13	0,26	637
		2 2	-0,06		0,09	***	0,25	0,00	-4,69	0,00	-1,16	0,25	637
Mortgage	*CAAR*	0 0	-0,01%		-0,01%		0,14	0,05	-0,07	0,94	-0,41	0,68	98
		1 1	0,81%		-0,51%		0,20	0,00	-0,03	0,98	0,69	0,49	98
		2 2	1,93%		1,09%		0,14	0,04	-0,75	0,45	1,16	0,25	98
	CAAB	0 0	-0,06		0,05	***	0,32	0,00	-2,51	0,01	-1,35	0,18	98
		1 1	-0,19		0,15	**	0,32	0,00	-2,41	0,02	-1,36	0,18	98
		2 2	-0,32		0,24	***	0,33	0,00	-2,46	0,01	-1,35	0,18	98

***, **, and * indicate statistical significance at the 1%, 5%, and 10% levels, respectively. *CAAR*: cumulative average abnormal return, *CAAB*: cumulative average abnormal beta shift, KS test: Kolmogorov-Smirnov test - p-value ≤ 0.05 indicates normal distribution at the 5% significance level, Wx-test: Wilcoxon signed rank test as non-parametric median test, [1]t-test: Parametric mean tests: Boehmer test for *CAARs*; Student's t-test for *CAABs*.

As Table 5.4 shows we indeed observe negative return effects and stable risk levels for Equity REITs, in line with the results from the total sample. Mortgage REITs show ambiguous return effects, without a statistical signifi-cance. The medians display a significant positive abnormal shift in risk. Nonetheless, the Kolmogorov-Smirnov test indicates, that the *CABs* are normally distributes and therefore the means and the respective t-values should be considered when evaluating the significance of risk shifts. The means in contrast to the medians do not show any significant risk shifts compared to general equity market behavior.

The observations from the recent analyses lead us to the general interpretation that Equity REITs as hard assets generally suffer compared to common equities. This might to some degree be caused by the moral hazard effect bank bailouts induce, favoring an appetite for risk, which is visibly carried over to the investment behavior. Another factor could be that the market might react faster to the positive signaling effect of a bank bailout, and hence outperforms Equity REITs in the short run. These factors influence our expectations towards the REIT subsamples in the following way: We expect those sectors which in times of market uncertainty generally have proven to be the most stable, now suffer the most in the event of bank bailouts, while subsectors closer to the general market behavior should show no difference to the market or might even profit. See Ro and Ziobrowski (2011) and Payne (2006) for relevant previous evidence on the performance of property-focused REITs.

Industrial REITs: Investing in warehouse buildings, their steady cash flows are secured by long term leases in multiuser properties. Generally, their long term development relies on the economic development. Overall, industrial REITs should be a relatively stable REIT type, which in the case of bank bailouts leads us to expect a tendency toward a negative valuation compared to common equities

Lodging REITs: They generate their cash flows by investing in hotel properties that cater to business and leisure travelers. Travelling represents generally one among the first expense categories affected by budget restrictions during crises but certainly also rises strongly in boom times. Thus, Lodging REITs should earn higher returns at higher risk levels. Bank bailouts should favor Lodging REITs compared to other REITs, which should give them a rather neutral, if not positive valuation compared to the market. See also Jackson (2009) for a performance study on Lodging REITs.

Office REITs: Generally invested in commercial properties, often in downtown locations they obtain their income on the basis of medium-term leases of mostly multitenant buildings. Their proximity to the tertiary sector makes them strongly dependent on the economic development. On the other hand their property location close to urban centers should devalue relatively little during recessions due to high marketability and potential multifunctional and multi-tenant use. We see Office REITs as a rather hard asset class and therefore expect them to suffer from a relatively negative valuation in the event of bank bailouts.

Retail REITs: Investing in shopping centers and malls, their cash flow depends on household spending. As bank bailouts do not directly impact households, spending levels should not be effected significantly due to this type of event. Retail REITs are thus expected to decrease their performance relative to common equities. E.g., Zietz, Sirmans, and Friday (2003) provide a performance analysis on the more traditional REIT types such as Retail and Office REITs.

Residential REITs: Investing in rental apartment housing, they should be relatively distant to short-term market developments and hence should represent a rather stable type of REITs. This should penalize them in the event of bailouts and makes us expect a negative valuation compared to common equities. We recommend considering the study of Newell and Fischer (2009) that documents the role of Residential REITs in REIT portfolios. They identify, a.o., a generally beneficial status of Residential REITs caused by decreasing home ownership versus increased use of rental accommodation post to the subprime and financial crisis.

Healthcare REITs: Investing in seniors housing and medical offices, they receive income that depends mainly on insurance and retirement payments that should be recession resistant. Healthcare REITs should be one of most stable REIT types, in principle being the most distant from short-term economic developments. Therefore, bank bailouts should result in a negative valuation compared to the market which should be able to outperform this type of REIT in the short run after positive market signaling events.

Specialized REITs: Investing in smaller property segments such as public storage or timber, they rely on unique demand segments for their cash flows. Thus, Specialized REITs may operate at either low or high risk levels and experience either positive or negative valuation effects in relation to common equities. See also Newell and Peng (2006) for a performance analysis of non-traditional REIT types such as healthcare and specialty.

Diversified REITs: Through their diversified investment style, both regarding property types as well as equity and mortgage investment, Diversified REITs should generally represent the common equity market. Therefore, we do not expect any abnormal risk or return shifts.

As Table 5.5 (next page) shows the expectation for significantly negative *CAARs* is largely supported. For the event date itself hardly any abnormal returns are observable but for the event windows [-1,+1] as well as for [-2,+2] we observe significantly negative *CAARs* for Office, Lodging, Retail, Residential and Healthcare REITs. Industrial and Diversified REITs show now significant *CAARs*. Specialized REITs show significant positive *CAARs*, but only on the event date itself. Regarding the analysis of the risk part, the picture is even more ambiguous. We observe significant positive risk shifts for Office, Residential, Specialized, and Diversified REITs. Industrial and Retail REITs show no significant risk shift, while Lodging and Healthcare REITs even show a significant negative risk shift.

Concluding, we find a clearly negative valuation effect (negative return and increased risk) for Office and Residential REITs which is in line with our expectations. For Retail, Specialized, and Diversified REITs we observe negative valuation effects (either negative return or positive risk shift). For Retail REITs this confirms our expectation. On the contrary, we did not expect negative valuation effects for Diversified REITs. Due to their variant character we did not have a clear forecast for the behavior of Specialized REITs. Against our expectation Industrial REITs do not show negative valuation effects, but instead do not show any significant difference from the market development, neither for return nor for the risk part. For Healthcare REITs we had negative valuation expectations and for Lodging REITs we presumed a rather positive behavior compared to the market. For both REIT types the expectations cannot be confirmed nor denied. Both REIT types show negative return behavior which though is also accompanied by decreased risk levels and it cannot be determined which effect dominates.

Table 5.5: Risk/Return Event Study: REIT-type Subsamples (II/II)

REIT Type		[-x; +y]		Mean		Median		KS-test	p	Wx-test	p	t-test[1]	p	N
Industrial		0	0	0,47%		-0,13%		0,08	0,68	-0,17	0,87	0,37	0,71	70
	CAAR	1	1	-1,30%		-2,74%	*	0,10	0,40	-1,72	0,09	-0,95	0,34	70
		2	2	0,32%		-2,53%		0,13	0,17	-0,78	0,43	0,26	0,79	70
		0	0	0,01		0,03		0,19	0,01	-1,44	0,15	0,35	0,72	70
	CAAB	1	1	0,02		0,09		0,19	0,01	-1,41	0,16	0,36	0,72	70
		2	2	0,05		0,17		0,19	0,01	-1,43	0,15	0,44	0,66	70
Office		0	0	0,89%		0,24%		0,14	0,06	-0,89	0,37	1,54	0,12	84
	CAAR	1	1	-2,51%	***	-2,98%	***	0,09	0,45	-2,65	0,01	-2,55	0,01	84
		2	2	-2,70%	***	-2,46%	**	0,08	0,57	-2,10	0,04	-2,49	0,01	84
		0	0	0,03	**	0,02	***	0,14	0,07	-2,88	0,00	2,45	0,02	84
	CAAB	1	1	0,08	***	0,05	***	0,13	0,11	-2,66	0,01	2,51	0,01	84
		2	2	0,13	**	0,08	***	0,13	0,11	-2,60	0,01	2,45	0,02	84
Lodging		0	0	-0,59%		-0,86%		0,13	0,13	-0,38	0,70	-0,93	0,35	77
	CAAR	1	1	-3,34%	***	-3,59%	***	0,07	0,87	-2,51	0,01	-2,89	0,00	77
		2	2	-2,55%	**	-0,88%		0,14	0,08	-1,19	0,24	-2,10	0,04	77
		0	0	-0,10	*	-0,01		0,26	0,00	-0,84	0,40	-1,79	0,08	77
	CAAB	1	1	-0,31	*	-0,02		0,26	0,00	-0,78	0,43	-1,80	0,08	77
		2	2	-0,51	*	-0,04		0,26	0,00	-0,76	0,44	-1,80	0,08	77
Retail		0	0	0,79%		-0,52%		0,12	0,02	-0,76	0,45	1,28	0,20	154
	CAAR	1	1	-1,61%	**	-1,52%	***	0,09	0,13	-2,45	0,01	-2,38	0,02	154
		2	2	-2,44%	***	-1,74%	**	0,10	0,07	-2,41	0,02	-2,53	0,01	154
		0	0	0,00		0,01		0,18	0,00	-1,02	0,31	-0,34	0,73	154
	CAAB	1	1	-0,02		0,03		0,18	0,00	-0,88	0,38	-0,40	0,69	154
		2	2	-0,03		0,04		0,18	0,00	-0,80	0,42	-0,42	0,67	154
Residential		0	0	-0,57%		-0,79%		0,11	0,19	-1,51	0,13	-1,13	0,26	98
	CAAR	1	1	-3,12%	***	-3,30%	***	0,09	0,35	-4,02	0,00	-3,98	0,00	98
		2	2	-4,11%	***	-4,34%	***	0,09	0,42	-5,05	0,00	-5,01	0,00	98
		0	0	0,03	***	0,02	***	0,16	0,01	-3,97	0,00	3,45	0,00	98
	CAAB	1	1	0,09	***	0,06	***	0,16	0,01	-3,93	0,00	3,59	0,00	98
		2	2	0,16	***	0,10	***	0,16	0,01	-3,93	0,00	3,59	0,00	98
Healthcare		0	0	0,71%	**	0,19%		0,07	0,84	-1,40	0,16	2,02	0,04	84
	CAAR	1	1	-1,94%	***	-0,88%	**	0,08	0,63	-2,21	0,03	-2,85	0,00	84
		2	2	-2,93%	***	-3,21%	***	0,06	0,89	-3,04	0,00	-3,17	0,00	84
		0	0	-0,06	*	0,03	*	0,35	0,00	-2,39	0,02	-1,74	0,09	84
	CAAB	1	1	-0,20	*	0,10	*	0,36	0,00	-2,31	0,02	-1,81	0,07	84
		2	2	-0,34	*	0,16	*	0,36	0,00	-2,25	0,02	-1,84	0,07	84
Specialized		0	0	1,47%	**	0,78%	*	0,19	0,01	-1,68	0,09	2,32	0,02	70
	CAAR	1	1	-0,70%		-0,79%		0,16	0,06	-1,52	0,13	-1,09	0,28	70
		2	2	-1,16%		-2,14%	**	0,21	0,00	-2,09	0,04	-1,29	0,20	70
		0	0	0,02	***	0,02	***	0,12	0,23	-3,18	0,00	2,72	0,01	70
	CAAB	1	1	0,07	***	0,05	***	0,13	0,16	-2,97	0,00	2,52	0,01	70
		2	2	0,11	**	0,09	***	0,13	0,20	-2,95	0,00	2,46	0,02	70
Diversified		0	0	0,17%		-0,33%		0,09	0,26	-0,30	0,76	0,34	0,73	112
	CAAR	1	1	0,92%		-0,49%		0,11	0,11	-0,09	0,93	0,39	0,69	112
		2	2	1,78%		0,99%		0,12	0,09	-0,53	0,60	0,79	0,43	112
		0	0	0,02	*	0,03	***	0,17	0,00	-3,45	0,00	1,65	0,10	112
	CAAB	1	1	0,07		0,08	***	0,16	0,00	-3,36	0,00	1,63	0,11	112
		2	2	0,12	*	0,14	***	0,15	0,01	-3,35	0,00	1,68	0,10	112

***, **, and * indicate statistical significance at the 1%, 5%, and 10% levels, respectively. *CAAR*: cumulative average abnormal return, *CAAB*: cumulative average abnormal beta shift, KS test: Kolmogorov-Smirnov test - p-value ≤ 0.05 indicates normal distribution at the 5% significance level, Wx-test: Wilcoxon signed rank test as non-parametric median test, [1]t-test: Parametric mean tests: Boehmer test for *CAARs*; Student's t-test for *CAABs*.

5.4.3. Bank Bailouts' Effect on Other Financial Stock Subsamples

To complete our insight on financial stock behavior we will now look into the other financial stock subsamples.

Asset Managers: On the one hand the business of asset managers it closely connected to the other companies of the financial sector, on the other hand the stable character of their revenues should give them a rather stable standing which may hinder their valuation in times of bank bailouts.

Banks: The findings of Hakenes and Schnabel (2009) show that the bailout of banks leads to increased risk levels of these banks' competitors. Klingebiel et al. (2001) state that bank returns do not react significantly to bailout programs. On basis of this research we expect negative valuation effects for banks in the event of bank bailouts, not driven by return effects but by increasing risk levels.

Table 5.6: Risk/Return Event Study: FSI-type Subsamples

FSI Type		[-x; +y]	Mean		Median		KS-test	p	Wx-test	p	t-test[1]	p	N
Banks	*CAAR*	0 0	-0,56%	***	-0,16%	***	0,10	0,00	-2,88	0,00	-3,71	0,00	833
		1 1	0,34%		0,21%		0,10	0,00	-0,46	0,64	0,56	0,58	833
		2 2	-0,10%		-0,14%		0,10	0,00	-1,31	0,19	-0,95	0,34	833
	CAAB	0 0	0,03	***	0,03	***	0,13	0,00	-9,74	0,00	6,70	0,00	833
		1 1	0,08	***	0,09	***	0,13	0,00	-9,74	0,00	6,97	0,00	833
		2 2	0,14	***	0,15	***	0,12	0,00	-9,79	0,00	7,16	0,00	833
Asset Mngrs	*CAAR*	0 0	-1,04%	**	-0,20%		0,15	0,00	-1,55	0,12	-2,36	0,02	210
		1 1	-0,97%		-0,56%	**	0,11	0,02	-1,97	0,05	-1,31	0,19	210
		2 2	-0,93%		0,29%		0,07	0,18	-0,71	0,48	-1,07	0,28	210
	CAAB	0 0	0,04	***	0,03	***	0,12	0,00	-6,48	0,00	5,29	0,00	210
		1 1	0,11	***	0,09	***	0,12	0,00	-6,42	0,00	5,12	0,00	210
		2 2	0,19	***	0,14	***	0,12	0,00	-6,38	0,00	5,10	0,00	210
Brokers	*CAAR*	0 0	-0,81%	*	-1,09%	***	0,18	0,00	-2,61	0,01	-1,87	0,06	133
		1 1	-0,35%		-0,09%		0,11	0,06	-1,14	0,25	-1,00	0,32	133
		2 2	-0,74%		-0,87%	**	0,13	0,03	-1,96	0,05	-1,40	0,16	133
	CAAB	0 0	0,02	**	0,02	***	0,08	0,34	-2,62	0,01	2,32	0,02	133
		1 1	0,05	**	0,05	***	0,08	0,38	-2,57	0,01	2,23	0,03	133
		2 2	0,09	**	0,09	***	0,08	0,38	-2,62	0,01	2,29	0,02	133
Insurances	*CAAR*	0 0	0,08%		0,22%		0,12	0,00	-1,32	0,19	-0,31	0,76	427
		1 1	-0,13%		-0,27%		0,13	0,00	-0,99	0,32	-1,21	0,23	427
		2 2	-0,23%		-0,11%		0,13	0,00	-0,52	0,60	-1,27	0,20	427
	CAAB	0 0	0,04	***	0,04	***	0,11	0,00	-9,87	0,00	8,34	0,00	427
		1 1	0,13	***	0,13	***	0,11	0,00	-9,85	0,00	8,30	0,00	427
		2 2	0,22	***	0,22	***	0,11	0,00	-9,81	0,00	8,32	0,00	427

***, **, and * indicate statistical significance at the 1%, 5%, and 10% levels, respectively. *CAAR*: cumulative average abnormal return, *CAAB*: cumulative average abnormal beta shift, KS test: Kolmogorov-Smirnov test - p-value ≤ 0.05 indicates normal distribution at the 5% significance level, Wx-test: Wilcoxon signed rank test as non-parametric median test, [1]t-test: Parametric mean tests: Boehmer test for *CAARs*; Student's t-test for *CAABs*.

Brokers: The operating business of brokers is generally closely connected to the general economic development. They should equally participate in the positive development triggered by the positive signaling effect of bank bailouts. Therefore, we do not expect any significant difference from the general markets' behavior.

Insurances: Through their diversified investment style, insurances should follow the general market development relatively closely. Nonetheless, their diversified income and cost structure should make them a rather stable equity class. Concluding, we expect insurances to show no significantly different behavior from the market, maybe with slightly negative valuation tendencies.

As Table 5.6 shows we find rare evidence for negative *CAARs* (Banks, Asset Manager, and Brokers each only at the event date itself). Nonetheless, we identify a negative valuation effect driven by increased risk levels which is clearly observable for all four subsamples as all event windows show significant positive risk shifts.

5.4.4. Robustness Analysis on Differences between Single Bailout Dates

Finally, we investigate whether the observed results are constant over time, or if we find diverging results for different bailout dates and if the USD value of bailed out assets has an influence on the behavior of REITs and other financial stock. Generally, we expect the results to be consistent over time. However, higher USD bailout volumes might have a stronger impact than smaller ones.

As Table 5.7 shows, in the case of REITs we find a significant negative return effect for four of the seven event dates, including the two dates with the highest amount of bailed out assets. Nonetheless, we also find significantly positive *CAARs* for two event dates. The remaining seventh event date shows only little evidence for a positive return effect. Throughout the seven bailout dates there is hardly any evidence for a risk shift. Hence, we mostly find evidence on a negative valuation effect for REITs in the event of bank bailouts supporting our findings from the total sample. The picture is not univocal though.

The analysis of the other financial stocks, as seen in Table 5.8, shows a similar picture for the return part. We find significant negative valuation effects for three of the seven event dates, including the dates one, two, and four by USD volume of the bailed out assets. Nonetheless, also for FSIs we find significantly positive *CAARs* for three of the event dates. One event date shows no significant difference in returns from the market. Comparing the result of the return event studies for REITs and FSIs, we find that for the five largest bailouts both groups show similar results in direction and significance of the results. For the risk part however, we observe significantly positive risk shifts for the six largest bailout dates for FSIs.

Table 5.7: Risk/Return Event Study: Date Subsamples and Total Sample REITs

Date and USD amount of bailed out assets		[-x; +y]	Mean		Median		KS-test	p	Wx-test	p	t-test[1]	p	N
10/28/2008 USD 115 bil.	*CAAR*	0 0	-2,05%	***	-1,05%	***	0,10	0,16	-2,64	0,01	-2,67	0,01	121
		1 1	-1,84%	**	-2,14%	***	0,10	0,13	-3,09	0,00	-2,43	0,02	121
		2 2	-1,38%	*	-2,15%	**	0,13	0,03	-2,21	0,03	-1,78	0,08	121
	CAAB	0 0	0,04	*	0,04	***	0,18	0,00	-3,33	0,00	1,70	0,09	121
		1 1	0,10		0,11	***	0,18	0,00	-2,66	0,01	1,41	0,16	121
		2 2	0,15		0,18	***	0,18	0,00	-2,56	0,01	1,34	0,18	121
11/17/2008 USD 33 bil.	*CAAR*	0 0	-0,91%	**	-0,62%	*	0,13	0,03	-1,62	0,10	-2,16	0,03	121
		1 1	-8,86%	***	-7,93%	***	0,13	0,04	-7,96	0,00	-9,55	0,00	121
		2 2	-12,56%	***	-9,87%	***	0,09	0,21	-8,55	0,00	-10,24	0,00	121
	CAAB	0 0	0,00		0,02		0,20	0,00	-1,52	0,13	-0,08	0,93	121
		1 1	0,00		0,06		0,21	0,00	-1,50	0,13	-0,05	0,96	121
		2 2	0,01		0,11		0,20	0,00	-1,62	0,11	0,11	0,91	121
12/31/2008 USD 15 bil.	*CAAR*	0 0	3,82%	***	2,48%	***	0,16	0,00	-7,58	0,00	8,07	0,00	121
		1 1	4,99%	***	3,99%	***	0,16	0,01	-7,48	0,00	8,59	0,00	121
		2 2	-3,05%	***	-5,94%	***	0,14	0,01	-4,36	0,00	-3,29	0,00	121
	CAAB	0 0	-0,05		0,02	*	0,32	0,00	-1,67	0,09	-1,59	0,12	121
		1 1	-0,15		0,06	*	0,32	0,00	-1,67	0,10	-1,59	0,12	121
		2 2	-0,26		0,09		0,32	0,00	-1,42	0,16	-1,63	0,11	121
01/09/2009 USD 14 bil.	*CAAR*	0 0	-1,29%	***	-1,85%	***	0,14	0,02	-5,12	0,00	-5,42	0,00	121
		1 1	-3,64%	***	-4,93%	***	0,12	0,06	-5,93	0,00	-6,90	0,00	121
		2 2	0,08%		-1,39%		0,09	0,25	-0,80	0,42	-0,63	0,53	121
	CAAB	0 0	-0,04	**	0,00		0,27	0,00	-0,68	0,50	-1,98	0,05	121
		1 1	-0,11	**	0,01		0,28	0,00	-0,71	0,48	-1,98	0,05	121
		2 2	-0,18	**	0,01		0,28	0,00	-0,69	0,49	-1,97	0,05	121
12/19/2008 USD 2.8 bil.	*CAAR*	0 0	4,08%	***	3,35%	***	0,08	0,36	-5,74	0,00	6,14	0,00	121
		1 1	1,57%	*	0,88%	*	0,12	0,04	-1,89	0,06	1,75	0,08	121
		2 2	4,49%	***	3,14%	***	0,10	0,16	-4,40	0,00	4,53	0,00	121
	CAAB	0 0	-0,01		0,06	***	0,28	0,00	-3,75	0,00	-0,42	0,67	121
		1 1	-0,03		0,18	***	0,29	0,00	-3,87	0,00	-0,35	0,72	121
		2 2	-0,05		0,29	***	0,28	0,00	-3,81	0,00	-0,36	0,72	121
12/23/2008 USD 1.9 bil.	*CAAR*	0 0	0,54%		0,09%		0,11	0,12	-0,66	0,51	1,16	0,25	121
		1 1	0,19%		-0,67%		0,14	0,02	-0,98	0,33	0,34	0,73	121
		2 2	4,26%	***	2,17%	***	0,15	0,01	-4,35	0,00	3,98	0,00	121
	CAAB	0 0	-0,01		0,06	***	0,28	0,00	-3,88	0,00	-0,20	0,85	121
		1 1	-0,02		0,18	***	0,28	0,00	-3,91	0,00	-0,17	0,87	121
		2 2	-0,02		0,30	***	0,28	0,00	-3,95	0,00	-0,16	0,87	121
03/13/2009 USD 1.45 bil.	*CAAR*	0 0	-1,65%	***	-2,43%	***	0,14	0,01	-4,77	0,00	-3,11	0,00	121
		1 1	-1,84%	**	-5,07%	***	0,15	0,01	-3,34	0,00	-1,94	0,05	121
		2 2	-1,09%		-3,80%	***	0,15	0,01	-2,58	0,01	-1,12	0,26	121
	CAAB	0 0	-0,02		0,01	***	0,33	0,00	-2,48	0,01	-1,02	0,31	121
		1 1	-0,05		0,03	**	0,33	0,00	-2,29	0,02	-1,04	0,30	121
		2 2	-0,08		0,04	**	0,33	0,00	-2,27	0,02	-1,05	0,30	121

***, **, and * indicate statistical significance at the 1%, 5%, and 10% levels, respectively. *CAAR*: cumulative average abnormal return, *CAAB*: cumulative average abnormal beta shift, KS test: Kolmogorov-Smirnov test - p-value ≤ 0.05 indicates normal distribution at the 5% significance level, Wx-test: Wilcoxon signed rank test as non-parametric median test, [1]t-test: Parametric mean tests: Boehmer test for *CAARs*; Student's t-test for *CAABs*.

Table 5.8: Risk/Return Event Study: Date Subsamples and Total Sample FSIs

Date and USD amount of bailed out assets		[-x ; +y]	Mean		Median		KS-test	p	Wx-test	p	t-test[1]	p	N
10/28/2008 USD 115 bil.	*CAAR*	0 0	-4,29%	***	-2,87%	***	0,07	0,16	-7,51	0,00	-8,88	0,00	229
		1 1	-3,15%	***	-2,54%	***	0,06	0,44	-5,48	0,00	-6,29	0,00	229
		2 2	-2,48%	***	-1,17%	***	0,08	0,08	-3,58	0,00	-4,06	0,00	229
	CAAB	0 0	0,06	***	0,03	***	0,11	0,01	-4,54	0,00	5,09	0,00	229
		1 1	0,15	***	0,09	***	0,11	0,01	-4,22	0,00	4,87	0,00	229
		2 2	0,24	***	0,13	***	0,11	0,01	-4,12	0,00	4,78	0,00	229
11/17/2008 USD 33 bil.	*CAAR*	0 0	-0,15%		0,45%		0,11	0,01	-0,86	0,39	-0,37	0,71	229
		1 1	-2,21%	***	-1,39%	***	0,08	0,11	-4,51	0,00	-4,88	0,00	229
		2 2	-4,52%	***	-2,31%	***	0,11	0,01	-5,87	0,00	-6,51	0,00	229
	CAAB	0 0	0,03	***	0,01	***	0,10	0,02	-2,99	0,00	3,39	0,00	229
		1 1	0,08	***	0,04	***	0,10	0,02	-2,87	0,00	3,23	0,00	229
		2 2	0,14	***	0,08	***	0,10	0,03	-3,06	0,00	3,34	0,00	229
12/31/2008 USD 15 bil.	*CAAR*	0 0	1,92%	***	1,30%	***	0,16	0,00	-6,29	0,00	5,61	0,00	229
		1 1	2,80%	***	1,84%	***	0,14	0,00	-6,58	0,00	6,53	0,00	229
		2 2	-0,08%		-1,46%		0,10	0,01	-1,50	0,13	-0,61	0,54	229
	CAAB	0 0	0,03	***	0,04	***	0,18	0,00	-6,33	0,00	3,71	0,00	229
		1 1	0,08	***	0,13	***	0,18	0,00	-6,33	0,00	3,68	0,00	229
		2 2	0,13	***	0,21	***	0,17	0,00	-6,33	0,00	3,78	0,00	229
01/09/2009 USD 14 bil.	*CAAR*	0 0	-1,40%	***	-1,14%	***	0,09	0,06	-6,48	0,00	-6,81	0,00	229
		1 1	-2,43%	***	-2,28%	***	0,06	0,34	-6,33	0,00	-7,00	0,00	229
		2 2	-1,96%	***	-1,38%	***	0,07	0,17	-4,02	0,00	-4,35	0,00	229
	CAAB	0 0	0,02	***	0,03	***	0,17	0,00	-6,39	0,00	3,27	0,00	229
		1 1	0,06	***	0,08	***	0,16	0,00	-6,60	0,00	3,63	0,00	229
		2 2	0,09	***	0,13	***	0,17	0,00	-6,63	0,00	3,68	0,00	229
12/19/2008 USD 2.8 bil.	*CAAR*	0 0	0,22%		0,01%		0,14	0,00	-0,59	0,55	0,50	0,62	229
		1 1	1,32%	***	1,43%	***	0,11	0,01	-3,69	0,00	3,20	0,00	229
		2 2	1,66%	***	1,28%	***	0,10	0,02	-3,46	0,00	3,27	0,00	229
	CAAB	0 0	0,05	***	0,06	***	0,08	0,07	-8,77	0,00	9,18	0,00	229
		1 1	0,15	***	0,17	***	0,09	0,05	-8,72	0,00	9,03	0,00	229
		2 2	0,26	***	0,30	***	0,09	0,04	-8,71	0,00	9,01	0,00	229
12/23/2008 USD 1.9 bil.	*CAAR*	0 0	-0,04%		0,14%		0,14	0,00	-0,01	0,99	-0,03	0,98	229
		1 1	0,21%		-0,20%		0,11	0,01	-0,08	0,93	0,74	0,46	229
		2 2	0,37%		0,46%		0,11	0,01	-1,06	0,29	0,98	0,33	229
	CAAB	0 0	0,05	***	0,06	***	0,11	0,01	-8,47	0,00	8,40	0,00	229
		1 1	0,16	***	0,19	***	0,11	0,01	-8,52	0,00	8,42	0,00	229
		2 2	0,26	***	0,31	***	0,11	0,01	-8,57	0,00	8,63	0,00	229
03/13/2009 USD 1.45 bil.	*CAAR*	0 0	0,43%		-0,03%		0,11	0,01	-0,08	0,94	1,14	0,25	229
		1 1	3,37%	***	1,45%	***	0,16	0,00	-3,67	0,00	4,32	0,00	229
		2 2	4,91%	***	2,90%	***	0,14	0,00	-5,56	0,00	5,82	0,00	229
	CAAB	0 0	0,00		0,01	***	0,23	0,00	-3,64	0,00	0,16	0,87	229
		1 1	0,01		0,04	***	0,22	0,00	-3,79	0,00	0,34	0,74	229
		2 2	0,01		0,07	***	0,21	0,00	-3,69	0,00	0,34	0,73	229

***, **, and * indicate statistical significance at the 1%, 5%, and 10% levels, respectively. *CAAR*: cumulative average abnormal return, *CAAB*: cumulative average abnormal beta shift, KS test: Kolmogorov-Smirnov test - p-value ≤ 0.05 indicates normal distribution at the 5% significance level, Wx-test: Wilcoxon signed rank test as non-parametric median test, [1]t-test: Parametric mean tests: Boehmer test for *CAARs*; Student's t-test for *CAABs*.

Generally speaking, our findings from the total sample analysis are validated. Nonetheless, we do not observe a completely univocal picture concerning the return analysis across the different event dates.

5.5. Conclusion

This research examines whether REITs, as financial stock, react differently from common equities when a monetary shock to the market occurs through the injection of liquidity by the government in form of bank bailouts. The objective is to advance previous research on the behavior of REITs in the case of monetary shocks. Previous research has shown that REITs react similar to common equities in the event of monetary shocks through a change in federal fund rates. We believe though, that the type of monetary shock is of crucial importance on how it affects REIT behavior and extend our previous evidence on bank failures. Bank failures have shown to cause positive valuation effects compared to common equities. Although bank failures and federal fund rate increases both trigger a limitation in market liquidity, they provoke opposing valuation effects for REITs. This might to some degree be attributable to the fact that they generally occur in completely different market situations. Since bank bailouts as well as federal fund rate decreases generally occur in similar market phases, as to speak in a recession, our null hypothesis states that REITs risk-return profile should not behave significantly different from other equities in the case of bank bailouts.

However, our empirical results for the total sample show that REITs generally experience significant negative abnormal returns (between -0.25% and -1.35%) around bank bailout dates, while we do not observe a shift in systematic risk levels. These results are in line with the result of our previous evidence on bank failures which showed positive return effects around bank failures, while risk levels remained stable.

Comparing the results on REITs with the total sample of other FSIs, we find comparable results with regard to the return analysis, however, in contrast to REITs, FSIs show a significant increase in risk levels. Generally speaking, REITs, as well as the other financial stocks, show a negative valuation effect around bank bailouts. This might bias our findings, as we might not be observing a unique behavior of REITs, but of the entire financial stock sector.

The analysis of the REIT subsamples provides us with valuable insights on how diverse the different REITs types react to the monetary shock of bank bailouts. Generally, the findings on the subsamples add well to our evidence from previous research on bank failures, as almost all REIT classes show reverse valuation effects comparing bank bailouts and bank failures. The analysis of other FSIs shows, that the negative return effect is only clearly observable for two of the four subsamples, while a positive risk shift is clearly displayed on a highly significant level throughout all subsamples.

Concluding, we find that the effect monetary shocks have on REITs, compared to the general stock market, depends highly on the type of monetary shock. Bank bailouts and bank failures have reverse valuation effects on REIT risk-return profile. Their counterparts from a market liquidity point of view, decreases and increases in federal fund rates also oppose their effects. In our opinion we find evidence supporting the safe heaven theory of Mueller and Mueller (2003) indentifying foremost Equity REITs as hard asset in times of market uncertainty, in that degree, that these REITs are outperformed by the market around the positive market signaling effect of bank bailouts. Furthermore, we find increasing risk levels for financial services institutions which might be a sign for the moral hazard problem involved with bailouts.

As our research has shown comparable results for REITs and other financial stocks it is of high interest to scholars of financial and real estate economics to further investigate interdependencies of REITs as financial stock and other industries that also belong to this group, in order to verify what characteristics are generally attributable to financial stock and which ones are REIT-specific.

The major limitations to our study are the following: First, we only research U.S. American REIT and financial stock market data, which might limit the explanatory power of this research. Secondly, although we took into consideration the time period from January 1999 until July 2010, our research is limited to a rather short period of time, as bank bailouts only occurred between January 2008 and July 2010. A third aspect concerns the fact, that after eliminating bailout dates that overlap with bank failures only seven event dates remain. We try to compensate this by conducting the analysis on single REIT level in order to investigate a sufficient amount of data points.

6. Banking Crisis and Government Intervention: How Bailouts versus Failures Affect Financial Stocks[13]

6.1. Introduction

The financial crisis starting in 2007 was primarily induced by the financial services industry (FSI). This industry does usually not only comprise banks, insurances, brokers, and asset managers but also the real estate sector. Speculative growth in real estate, not least in the subprime sector, had built up innumerable portfolios of mortgage backed securities traded among banks and other financial services institutions. Once the uncertainty about the creditworthiness of the underlying borrowers became prevalent, strongly invested banks slid into a severe solvency crisis (see Demyanyk and Hemert, 2011; Harrington, 2009; and Reinhard and Rogoff, 2008; 2009).

The bankruptcy of Lehman Brothers on September 15th 2008 constitutes a climax of this financial crisis as it showed that bank failures would be highly contagious in the strongly integrated financial system (see also Iyer and Peydró 2011). To prevent a cascade of bank failures and by this a potential meltdown of the total system the U.S. government decided to stop the vicious circle by announcing the Troubled Asset Relief Program (TARP). The intention of this bailout program was to calm down financial markets and prevent fire sales that would drive asset values far below reasonable measures and increase the risk of a strong and near-by spillover effect on the real economy (see Paulson, 2008 and Carlson, King, and Lewis, 2011).

However, two questions arise: First, did policy makers, who where continuously overwhelmed by the rapidness of financial markets, appropriately achieve stability of the financial services industry and second, did bailouts evoke positive wealth effects benefiting FSI shareholders. The objective of our study is thus to shed light on potential adjustments in banks' and other financial services institutions' risk-return profile through government bailouts. For robustness we also control for bank failures. Already Acharya et al. (2011) discuss how financial intermediaries and their capital have been affected by the crisis and thus prove the genuine interest into this industry. Our results allow an even broader perspective on evaluating government intervention as we illustrate how the

---

[13] I am grateful to Dirk Schiereck and Andreas Pfnür for valuable comments and insights.

FSI's performance itself responded to bailouts versus failures during the financial crisis.

In our analysis we concentrate on both first and second moments of mean returns analyzing the impact of seven major bank bailouts versus seven major bank failures on market, industry, and single stock performance. As the rapid progress of the financial crisis led to a high frequency of bailout and failure dates in short time periods we decide to only focus on short-term effects from two days before to two days after the events. Event study methodology (see Corrado, 2011) is applied to scrutinize for abnormal mean returns. We also observe simple average mean returns around the events as reference measure. Return risk behavior is primarily measured by applying an EGARCH (1,1) model (see Engle and Ng, 1993 and Nelson, 1991) to observe daily changes in variance estimates. We complement the risk perspective by also analyzing the impact on asset-to-market correlation which then enables us to determine short-term changes in the systematic risk estimator beta.

Overall we find zero average returns around bailout dates clearly distinguishing from strongly negative average returns around failure dates. Furthermore, our subsequent event study analysis returns ambiguous mean return effects of financial stocks around government intervention. While banks show two-sided significant abnormal returns around bank bailouts, the FSI subsector REITs shows strongly negative abnormal returns, opposed by insurances showing positive abnormal returns. For bank failures we find reverse mean abnormal returns.

Turning to the risk perspective on financial stocks, EGARCH variance estimates show a strict decrease around bailouts. Again this opposes financial stocks' reaction to bank failures with variance estimates rather increasing. Nevertheless the decreased stock variances around bailouts stabilize the system in the short run, increasing correlation coefficients finally lead to rising beta risk measures. Hence, short-term stability is paralleled by increased systematic risk loaded to the financial sector.

We show that in the short run the FSI subsectors are affected much stronger by bank bailouts and bank failures in both, first and second moments of mean returns than other industries we control for. However, we observe no symmetric FSI sub-sector behavior. We find REITs deviate significantly from other FSI's mean return behavior which leads to a compensation of abnormal returns for this asset class. One similarity across all analyzed asset classes is the observed *calm-before-the-storm* effect before bank events that could be shown due to the application of the EGARCH-method. We interpret this as evidence that bank failures and bank bailouts are well anticipated – failures then lead to increased uncertainty, whereas bailouts calm down general market volatility.

Even though lower market volatility around bailouts indicates enhanced system stability, policy makers should be concerned by the overall increase in systematic risk as a sign for moral hazard. Government intervention obviously leads to an increase in asset-to-market correlation which decreases risk diversification in the economy. Contrary, the beta risk decent around bank failures, especially for banks, shows that in the aftermath of excessive risk taking market participants decrease their risk appetite if the market functions autonomously. Flannery (2001) yet discusses the theory of "market discipline" and shows that it can actually incentive banks to control and if necessary reduce their risk exposure. Concluding, policy makers performing bailouts do achieve an immediate cool down of the system without generating exceptional wealth for financial stockholders. But, the price seems to be the creation of increased systematic risk that looms over the economy as a whole and by this augments its vulnerability to future crises.

The paper is structured as follows: In part 6.2 we review the literature; in part 6.3 we explain our data selection and the empirical methodology; in part 6.4 we discuss our results; part 6.5 concludes.

6.2. Literature Review

6.2.1. Financial Crisis and Government Intervention

We start with a review of timely literature on financial crises and their effects on asset valuations and the real economy. This is followed by evidence on Federal Reserve policy in times of crisis and how adjustments in federal fund rates or other regulatory measures affect financial intermediaries and common equity markets. If conventional measures are not successful bank liquidity crises can finally result in failures unless bailouts are undertaken. Both key-events have their own implications on the progress of a financial crisis.

Reinhart and Rogoff (2009) find that financial crises usually share three characteristics: First, equity markets declines are strong and persistent averaging around 55 percent throughout a recession of about three and a half years. Second, banking crises lead to deep declines in output (on average nine percent over two years) and employment (on average seven percent over four years). And third, the real value of government debt rises enormously as collapses in tax revenues add to bailout programs and countercyclical fiscal policies.

Acharya et al. (2009) analyze causes and remedies of the 2007-2009 financial crisis concluding, that the cost of integrating global financial markets is increased systemic fragility. They identify the collapse of two highly levered Bear Stearns hedge funds invested in asset-backed securities (ABS) as the most identifiable event of a typical financial crisis being a credit boom and an asset bubble. When collateral assets lose value, margin calls initiate fire sales, resulting in a vicious circle of declining asset prices.

Cecchetti (2009) discusses the difficult situation of the Federal Reserve during the 2007 crisis where banks' exposure to toxic ABS could have lead to a profound contraction in bank lending and a consecutive deep recession. Eroding bank capital due to losses exceeds standard central bank tools to provide liquidity. The need of banks for new equity capital from the outside, either by investors or through government bailouts, urged the Fed to take less-traditional actions such as making loans to individual borrowers.[14] A certain fiscal flavor may be interpreted as it could lead to subsidizing banks or even bailing them out.

Bernanke and Kuttner (2005) concentrate on a traditional tool of Federal Reserve policy and analyze the effect of federal fund rate adjustments on equity prices. They find that an unanticipated 25-basis-points cut leads to a 1% increase in the stock market index. The surprise fraction of a rate adjustment accounts for the major part in stock price reaction. Furthermore, they identify event induced variance as equity returns show a 10% increase in volatility on the monetary policy announcement dates. Interestingly, they find reactions showing industry dependence with high-tech and telecommunications appearing strongest versus energy and utilities indicating no significant responses. Financial services are not scrutinized individually.

Bomfim (2003) examines pre-announcement and news effects of federal fund rate adjustments on stock market volatility. Applying GARCH methodology he identifies abnormally low volatility on days before regularly policy announcements also called a *calm-before-the-storm* effect. Furthermore, an asymmetry in volatility response is uncovered as higher-than-expected increases in federal fund rates cause larger volatility effects than comparable decreases. Bredin, O'Reilly, and Stevenson (2007) extend this previous evidence by focusing on the asset class of REITs as financial stocks. Albeit the shock is significant in returns and volatility, they do not find evidence for asymmetry in volatility reaction nor a *calm-before-the-storm* effect.

Most recent evidence on potential contagion due to interbank linkages is provided by Iyer and Peydró (2011). They find that, as interbank lending represents a considerable part of banks' liability and borrowing sides, bank failure shocks are transmitted rapidly. Banks can react by hoarding liquidity leading

---

[14] The Fed's concern for system-wide financial stability let them invoke Article 13(3) of the Federal Reserve Act. By this, the Board of Governors can authorize to make loans to individual borrowers that are unable to obtain credit from a banking institution. For a further discussion of the lender of last resort see also Kaufman (1991).

eventually to a collapse of the interbank liquidity market as has happened within the recent financial crisis. Nevertheless, interbank contagion can still materialize which increases the probability of a true banking crisis. For further evidence on bank contagion see also Haldane and May (2011), Diamond and Rajan (2005), and Kaufman (1994).

Carlson, King, and Lewis (2011) analyze the relationship between the banking sector and the real economy concentrating on the implications of distress in the financial sector. When banks' capital bases erode they raise loan rates and retain funds by increasing deposit rates, resulting in a decline of borrowing activity and a restrain of economic growth. The economists construct an index named the Financial Institution Health Index (FIHI) showing that its deterioration incurs a tightening of lending standards. They also identify a rather strong relationship to the subsequent development of economic growth. See Dell'Ariccia, Detragiache, and Rajan (2008) for comparable evidence on spillover effects to the real economy, especially on sectors that are more dependent on external financing.

Gropp, Hakenes, and Schnabel (2011) investigate the competitive effects of bank bailout policies on the banking sector and find that guarantees lead to a strong increase in risk taking behavior of competitor banks. However, they do not find that banks that have been bailed out decrease their risk adversity themselves. Nevertheless, this supports moral hazard theory and yet Cordella and Yeyati (2003) have shown that ex-ante commitment by central banks to assist distressed financial institutions induces such. But, they also observe a risk-reducing value effect as bank failure frequency is reduced which actually overcompensates the moral hazard implications and results in a decreased bank risk.

Giannetti and Simonov (2009) find that the recapitalization of a financial intermediary leads to a positive value effect for its customers. Also Gorton and Huang (2004) report that market liquidity may effectively be restored and strengthened when banks are bailed out. So, the government is actually able to increase welfare when impeding a cascade of bank failures, even though moral hazard is provoked. Furthermore, Klingebiel et al. (2001) analyze stock market returns in response to bank restructuring and uncover that bank stocks as well as non-bank stocks react positively to government bailouts.

6.2.2. Contribution to the Literature and Basic Hypotheses

The literature review has shown the considerable interest of financial economists on the effects of banking crises onto the markets in general and the implications of its key-events, bank bailouts and bank failures, as special cases. Due to the fact that bailouts and failures considerably affect not only the liquidity of single entities but also many related counterparties, potentially the market as a whole, we interpret them in intuition comparably to federal fund rate adjustments by the Federal Reserve. Federal Reserve policy is yet a standard tool to affect economic growth in boom and recession times and its effect on risk and return profile of common equity markets so as REITs as financial stocks has yet been analyzed. However, current literature lacks evidence on how the key-events of a financial crisis, namely bank bailouts and bank failures, affect first and second moments of mean stock returns. We are eager to shed light on this gap and start by an analysis of the financial services sector as it is located in the centre of the financial crisis.

An analysis of the presented literature enables us to derive general arguments that serve as a framework for more precise hypotheses which are to be formulated in the respective results' sections. First, in the case of bank failures we may argument for a relatively strong negative return effect within the financial services sector compared to the effect on the real economy. The risk of immediate contagion within the financial sector is high but the effect on the real economy is expected to be somehow deferred. Second, considering return risk we may expect to observe a *calm-before-the-storm* effect as it could be observed before Federal Reserve policy announcements and market participants could also anticipate specific announcements regarding banks that find themselves in a severe liquidity crisis. Third, bank bailouts show to restore market confidence when announced but parallel they induce moral hazard. Hence we expect to observe both, a decrease in volatility but also an increase in some risk measure of, e.g., banks that are primarily affected by moral hazard reducing their risk adversity. This relates to point four, concerning our comparably ambiguous expectations on mean return behavior of financial stocks around bank bailouts. On the one hand bailouts split off the risk of failure potentially benefitting general valuations but on the contrary, government ownership in banks may lead to lower economic growth and reduced bank profitability in the long run (see, e.g., La Porta, Lopez-De-Silanes, and Shleifer, 2002).

6.3. Data and Methodology

6.3.1. Data Selection and Descriptive Statistics

As we observe the effect of U.S. bank failures and bank bailouts as key-events of the recent financial crisis on U.S. financial stocks and other industries our data sources are well diverse. Data on bank failures is obtained by the Federal Deposit Insurance Cooperation (FDIC, 2010) which provides a *Failed Bank List.*

Data on bank bailouts is provided by the U.S. Department of the Treasury (2011) listing the financial services institutions that where supported via the *Troubled Asset Relief Program* (TARP). Total return data in USD for the S&P 500 Composite as general market index, financial services indices, other industry indices and single financial stocks are obtained from Thomson Datastream.

In a previous study Raudszus, Olliges, and Mueller (2011) yet analyze the effect of 130 bank failures dates within the years 1999 to 2010 on REIT returns. In a consecutive study Olliges, Raudszus, and Mueller (2011) examine the opposite analyzing the effect of seven major bank bailout dates on REITs and other financial stocks. Both studies apply event study methodology to scrutinize for abnormal returns and beta aligned risk event studies to measure abnormal systematic risk shifts based on a GARCH (1,1) model. This previous evidence yet observes reverse valuation effects of bank failures and bank bailouts on REIT returns as financial stock.

We are eager to build upon these studies by taking a step back and investigating the behavior of the entire asset class of financial stocks with respect to both key-event-types of the recent financial crisis, failures and bailouts, at once. To eliminate any potential bias which may be induced by a largely different size of samples (amount of failure vs. bailout dates under analysis) we decide to contrast the seven bank bailouts with seven (major) bank failures.[15] Table 6.1 provides details on our financial crisis key-event date sample.

For our empirical analysis we consider total return data for a three year period from January 2007 to December 2010. The S&P 500 Composite serves as general market index and the FTSE NA Financials index represents financial stocks. The FTSE NA Industrials index represents the industrial goods industry (IGI), the FTSE NA Consumer Goods index represents the consumer goods industry (CGI) and the FTSE NA Consumer Services index represents the consumer services industry (CSI). All three are analyzed for robustness of the observed return and risk shifts in the financial sector. In a further analysis step we also use respective subindices for the above industry indices. These are the FTSE NA REIT, the FTSE NA Banks, the NA-DS Insurance, the NA-DS Asset Managers and the US-DS Insurance Brokers for the financial sector. For the industrials and the consumer goods sectors we analyze seven subindices each and

[15] We also contribute to the literature in that we advance and broaden our methodological approach measuring risk adjustments as we build upon an EGARCH (1,1) model. Methodological details will be elucidated some paragraphs below where we discuss our methodology as a whole.

for the consumer services industry we examine four subindices.[16] The risk and return behavior of the financial services subsectors is furthermore analyzed on a single stock level to increase statistical validity of results. Therefore, we analyze the return data of 119 single REITs, 119 banks, 61 insurance companies, 30 asset managers, and 19 broker agencies.[17]

Table 6.1: Event Date Sample Overview: Bank Failures and Bank Bailouts

Bank Failures				Bank Bailouts			
Event date	Day of week	Nr. of banks	Assets at default (USD bil.)	Event date	Day of week	Nr. of banks	Bailout capital (USD bil.)
07/11/2008	Monday	1	32.010	10/28/2008	Tuesday	8	115.000
07/25/2008	Friday	2	3.654	11/17/2008	Monday	21	33.561
09/15/2008[1]	Tuesday	1	16,000.000	12/19/2008	Friday	49	2.792
09/25/2008[2]	Thursday	1	307.000	12/23/2008	Tuesday	42	1.900
01/29/2010	Friday	6	5.531	12/31/2008	Wednesday	7	15.079
02/19/2010	Friday	4	4.171	01/09/2009	Friday	43	14.772
04/30/2010	Friday	7	25.825	03/13/2009	Friday	19	1.455

[1]) Failure of Lehman Brothers; [2]) Failure of Washington Mutual; Assets at default: cumulated bil. USD assets of respective banks that defaulted on a single date; Bailout capital: cumulated bil. USD bailout capital that was injected into respective banks, that where bailed out on a single date.

Table 6.2 provides descriptive statistics on mean return behavior of the market in general and the selected asset classes throughout the time period of the financial crisis under investigation. Overall, mean returns are statistically not different from zero and skewness is low. However, we observe typical stylized facts of financial return data such as fat tails and squared stock returns show strong serial correlation (Ljung-Box-test) and significant heteroscedasticity (ARCH-test).

---

[16] *(FTSE NA) Industrials:* Construction and Material (CON&MAT), Aero and Defense (AER&DEF), General Industrials (GEN IND), Electric Equipment (ELE EQU), Industrial Engineering (IND ENG), Industrial Transportation (IND TRA), Support Services (SUP SER); *Consumer Goods:* Auto and Parts (AUT&PAR), Beverages (BEVER), Food Production (FD PROD), Household Goods (HOU GDS), Leisure Goods (LEI GDS), Personal Goods (PER GDS), Tobacco (TABACCO); *Consumer Services:* Food and Drug Retailers (FD&DG RET), General Retailers (GEN RET), Media (MEDIA), Travel and Leisure (TRA&LEI).

[17] The REIT stock selection and respective return data is equivalent to Raudszus, Olliges, and Mueller (2011) and the data on other financial stocks is obtained as in Olliges, Raudszus, and Mueller (2011).

Table 6.2: Descriptive Statistics on Market and Industry Daily Returns

07/2008 - 07/2010[1]	S&P 500	FTSE NA Financials	FTSE NA Industrials	FTSE NA Cons. Gds	FTSE NA Cons. Svs.
Mean[2]	-0.03%	-0.05%	-0.03%	0.01%	0.01%
Median[2]	0.10%	0.00%	0.07%	0.06%	0.06%
Skewness	-0.15	-0.12	-0.25	0.07	0.07
Kurtosis	7.59	6.19	5.30	9.11	7.40
StD	0.02	0.04	0.02	0.02	0.02
Q_{10}	338[a]	216[a]	297[a]	310[a]	275[a]
Q_{100}	881[a]	901[a]	883[a]	694[a]	771[a]
$ARCH_{10}$	81[a]	44[a]	100[a]	65[a]	61[a]
$ARCH_{100}$	375[a]	168[a]	277[a]	389[a]	371[a]

[1]) Descriptive mean return statistics on the basis of 504 trading days throughout the period of the financial crisis under investigation; [2]) Mean and median returns are not significantly different from zero; Squared stock returns are tested by Ljung-Box test (see Ljung and Box, 1978) for serial correlation ($Q_{10/100}$) and by ARCH test (see Engle, 1982) for heteroscedasticity for 10 and 100 lags; [a]) indicates test result is positive for serial correlation or heteroscedasticity at the 5% significance level, [b]) indicates the opposite.

6.3.2. Methodology for Measuring Stock Return Risk

Next to the analysis of mean returns around bank failures and bailouts we will also scrutinize the behavior of return risk around these key-events. Return risk is primarily reflected by return variance but also by asset-to-market correlation and in combination of these by beta as systematic risk measure. Variance modeling is subject to a methodological discussion as multiple methods exist to obtain specific variance estimates.

Previous studies by Raudszus, Olliges, and Mueller (2011) and Olliges, Raudszus, and Mueller (2011) upon which we build yet apply a GARCH (1,1) model to measure daily variance adjustments. This standard GARCH model by Bollerslev (1990) yet incorporates one important stylized fact of financial stock data called volatility clustering which is the serial correlation of squared returns. It models variance as a partially autoregressive process. However, another prevalent stylized fact in stock return data is the asymmetry in volatility reaction to positive and negative return shocks. Hentschel (1995) provides a methodological overview on the GARCH-family and elucidates the so called leverage-effect, which describes that volatility increases more in respondence to negative return shocks as it does to comparable positive return shocks. Also Hansen and Lunde (2005), analyzing the forecasting performance of volatility models, find that a simple GARCH (1,1) is inferior to models that can accommodate a leverage effect. The exponential GARCH (EGARCH) by Nelson (1991) incorporates this asymmetry in its variance model as it determines whether a return series generally shows a leverage effect.

We are convinced that the basic EGARCH (1,1) model is appropriate for variance modeling in this study as previous evidence yet shows reverse abnormal return effects on REITs as financial stock. Previous studies though lack evidence on precise variance adjustments around these return shocks due to bank failures and bailouts. Applying EGARCH we first target to prove that financial stocks generally show a leverage effect. Second, we hope to observe reverse abnormal returns also for other financial stocks than REITs such as banks and insurances. And third, we are eager to find out if negative abnormal returns are followed by larger volatility increases relative to comparable positive abnormal returns.

EGARCH methodology has been widely applied in recent research. We build upon Ederington and Guan (2010) who analyze volatility adjustments of the S&P 500 with respect to return shocks applying among other methods an EGARCH (1,1) model for conditional volatility estimates. They find this asymmetric GARCH model to achieve a higher log-likelihood value than the standard GARCH (1,1) model describing time-series estimates of conditional volatility. In summary, they show negative return shocks have a large positive impact on volatility, zero and small positive returns have a negative impact on volatility, and large positive return shocks have again a positive but much weaker impact on volatility. Other recent studies applying EGARCH methodology are Adrian and Rosenberg (2008); McAleer, Chan, and Marinova (2007); Poon and Granger (2003); Darrat, Rahman, and Zhong (2003); and Chen and Kuan (2002).

The EGARCH (1,1) model to determine daily variance estimates, σ_t^2, takes the following form:

$$\log\sigma_t^2 = \gamma_1 + \gamma_2 \log\sigma_{t-1}^2 + \gamma_3\left[\frac{|\varepsilon_{t-1}|}{\sigma_{t-1}} - \sqrt{\frac{2}{\Pi}}\right] + \gamma_4\left(\frac{\varepsilon_{t-1}}{\sigma_{t-1}}\right) \tag{6.1}$$

Where γ_1 is a constant, γ_2 the coefficient for persistency in volatility, γ_3 the coefficient for the return shock *(ε)* absorption, and γ_4 the coefficient for the leverage effect.

Throughout the years 2007 to 2010 we estimate EGARCH (1,1) models for the S&P 500 Composite, the industry indices for financials, industrials, consumer goods and services, so as for the subindices of these industries and for all single stocks of the financial sector in our sample. Estimations of equation (6.1) are reported in Table 6.3 for the S&P and the industry indices. We find all coefficients to be highly significant and by this, a negative γ_4 clearly supports the existence of a leverage effect for each index. This does also hold true for the subindices (Table A6.1) and almost all single stocks. In the latter case Table A6.2 provides an overview of the single stock sample regarding EGARCH estimation which provides three exemplary companies per financial services subsector sorted by the magnitude of the leverage effect. The fact that some stocks do not show a leverage effect, in contrast to all indices showing it, is well established. Bekaert and Wu (2000) argue that firm returns are more correlated in downturns impacting index volatility positively as covariances rise, while single stock volatility stays unchanged.

Table 6.3: EGARCH (1,1) Estimation of Market and Industry Indices for 2007-2010

	S&P 500	FTSE NA Financials	FTSE NA Industrials	FTSE NA Consumer Goods	FTSE NA Consumer Services
γ_1	-0.20 ***	-0.14 ***	-0.14 ***	-0.26 ***	-0.20 ***
	-5.3	*-3.9*	*-3.6*	*-5.2*	*-4.3*
γ_2	0.98 ***	0.98 ***	0.98 ***	0.97 ***	0.98 ***
	212.9	*206.9*	*206.0*	*176.4*	*184.0*
γ_3	0.15 ***	0.24 ***	0.15 ***	0.14 ***	0.16 ***
	5.0	*7.2*	*5.7*	*4.9*	*5.5*
γ_4	-0.15 ***	-0.09 ***	-0.10 ***	-0.15 ***	-0.12 ***
	-6.6	*-4.9*	*-5.0*	*-6.4*	*-5.7*

γ_1 is a constant, γ_2 the coefficient for persistency in volatility, γ_3 the coefficient for return shock *(ε)* absorption, and γ_4 the coefficient for the leverage effect; t-values in parentheses; *** denotes significance at the 1% level.

Table 6.4: Descriptive Statistics on Market and Industry Daily EGARCH Variance Estimates

07/2008 - 07/2010[1]	S&P 500	FTSE NA Financials[2]	FTSE NA Industrials[2]	FTSE NA Consumer Goods	FTSE NA Consumer Services
Mean	.0004	.0012 ***	.0005 ***	.0002 ***	.0003 ***
Median	.0002	.0006 ***	.0003 ***	.0001 ***	.0002 ***
Skewness	2.28	1.40	1.75	2.62	2.37
Kurtosis	8.03	4.48	5.48	10.36	8.52
StD	.0005	.0012	.0005	.0003	.0004
Q_{10}	28[a]	19[a]	38[a]	20[a]	28[a]
Q_{100}	94[b]	85[b]	103[b]	82[b]	87[b]
$ARCH_{10}$	26[a]	27[b]	38[a]	13[b]	10[b]
$ARCH_{100}$	106[b]	80[a]	94[b]	64[b]	82[b]

[1]) Descriptive mean variance statistics on the basis of 504 trading days throughout the period of the financial crisis under investigation; [2]) FSI mean variance is also sig. different from Ind. mean variance at a 1% level; ***) standard t-test and Wilcoxon rank sum test find mean difference of single industries to the S&P500 at 1% significance level; Squared stock returns are tested by the Ljung-Box test (see Ljung and Box, 1978) for serial correlation ($Q_{10/100}$) and by ARCH test (see Engle, 1982) for heteroscedasticity for 10 and 100 lags; [a]) indicates test result is positive for serial correlation or heteroscedasticity at the 5% significance level, [b]) indicates the opposite.

Now, we can provide descriptive statistics on average variance estimates throughout the time period of the financial crisis under investigation (07/2008-07/2010). Table 6.4 provides evidence that financials' average variance rate is significantly higher than that of the general market and other industry variances. Average variances of the consumer goods and consumer services industries are even lower than that of the market. EGARCH estimation obviously leads to a rather skewed variance distribution and kurtosis equals about the one of the daily return distribution. Testing standardized squared returns, u^2/σ^2, for serial correlation and heteroscedasticity we find that both earlier prevalent characteristics in squared returns could be reduced strongly. This approves the appropriateness of applying an EGARCH model to estimate return variances in the present case.

Calculating simple one-year rolling asset-to-market correlation coefficients we obtain a further risk measure that is potentially subject to considerable adjustments throughout the financial crisis and around its key-events. Furthermore, daily variance and correlation estimates allow us to determine daily betas as systematic risk measure and by this a third stock return risk indicator. Table 6.5 provides a descriptive overview on correlation and beta characteristics throughout the two years under close investigation. We may report that correlation coefficients to the S&P 500 of selected non-financial industries are on average significantly higher than that of financial stocks. On the contrary, financial industry beta is on average significantly higher than that of its peers, which is driven by the yet observable higher average variance. Skewness is generally below an absolute value of one and below zero across industries for correlation estimates. Kurtosis measures are much closer around three compared to the return and variance distributions and thus indicate that correlation and beta estimates are closer to a normal distribution.

Table 6.5: Descriptive Statistics on Daily Industry-to-market Correlation and Beta Estimates

07/2008 - 07/2010[1]	FTSE NA Financials	FTSE NA Industrials	FTSE NA Consumer Goods	FTSE NA Consumer Services
Correlation to S&P				
Mean	0.90	0.96 ***	0.91 ***	0.94 ***
Median	0.90	0.96 ***	0.92 ***	0.94 ***
Skewness	-0.49	-0.02	-0.37	-0.60
Kurtosis	3.16	2.22	1.54	1.93
StD	0.01	0.01	0.03	0.02
Beta				
Mean	1.55	1.10 ***	0.65 ***	0.89 ***
Median	1.52	1.09 ***	0.65 ***	0.89 ***
Skewness	0.52	0.24	0.63	-0.21
Kurtosis	2.76	3.08	4.83	3.12
StD	0.32	0.11	0.08	0.09

[1]) Descriptive mean correlation statistics on the basis of 504 trading days throughout the period of the financial crisis under investigation; *** indicates mean difference significance at 1% level to the financials index applying the two-sided t-test and the Wilcoxon rank sum test.

Consult Table A6.3 for complete descriptive statistics on FSI subindices' daily returns, variances, correlations, and betas. Whereas mean returns are comparably around zero, we find first REITs and second bank average variances to be the highest among financial stocks. Again, the reduction in heteroscedasticity and serial correlation in standardized returns supports the appropriateness of applying EGARCH for estimating variance. Highest correlation to the S&P 500 show first, asset managers and second, insurers, but still REITs and then banks display highest beta estimates. Further descriptive statistics on the industrials, consumer goods, and services industries so as single financial stocks are not displayed but can be provided upon request.

6.3.3. Structure of the Return and Risk Analysis

The methodology for our combined return and risk analysis around bank failures and bank bailouts as key-events of the recent financial crisis consists of six consecutive steps. Our objective is to provide an extensive overview of return and risk behavior to obtain an exhaustive view on how the valuation of financial stocks evolves around bank events. The first five analyses observe return and risk adjustments from two trading days prior to two trading days post bank events. We concentrate on the immediate valuation effect of bailouts versus failures but also provide a view on potential information leakage or anticipation effects just before the bank events. The sixth analysis is a multivariate regression that aims to support and add to our then obtained insights on return and risk effects.

First, we observe average returns of the S&P 500 Composite, the FSI, and the IGI, CGI, and CSI industries around the key-events. Then, we also focus on FSI-subsector behavior individually by analyzing the same on a single stock level. Second, we perform event study methodology similar to Corrado (2011) to scrutinize for abnormal returns around failures and bailouts. We do this for the same data selection as before except for the S&P 500 because it serves as the general market index to the selected industry indices and the individual stocks. Third, we turn to stock return risk and measure respective shifts in the variance factor, σ^2. This is followed by a measurement of adjustments in the asset-to-market correlation coefficient, ρ_{im}, as fourth step. Last but not least, we determine the implications on beta as systematic risk factor applying Equation (6.2).

$$\beta_{it} = \frac{\rho_{im}\sigma_{it}}{\sigma_{mt}} \tag{6.2}$$

Multivariate regressions on FSI returns and variance estimates around bank failure and bailout dates are performed to identify potential drivers other than the industry.

6.4. Results and Discussion

6.4.1. Average Stock Returns around Bank Events

Our first analysis concentrates on standard average returns from two days prior to two days post bank failures versus bailouts. We perform this analysis to observe how the market's and selective industries' mean returns behave in absolute terms around these key-events. Furthermore, it facilitates the interpretation of abnormal returns as result of the subsequent analysis because it serves as a reference point.

As we assume bank failures liquidity shock to be comparable to that of federal fund rate increases we may expect comparative effects being generally negative mean returns. These should be strongly negative for the financial sector as it is immediately affected. The effect on the real economy should also be negative but less pronounced as it may suffer from anticipated spillover effects, discussed, e.g., by Calomiris (1995). Among financial stocks we expect banks to show strongest negative means as they constitute the primarily impacted subsector. As Raudszus, Olliges, and Mueller (2011) yet found positive abnormal returns for REITs around failures we expect this financial asset class to display least negative but still negative returns comparable to Bredin, O'Reilly, and Stevenson (2007).

Around bank bailouts we expect generally more positive mean returns compared to bank failures. Since, as we earlier declared, the capital injection in its effect to market liquidity is comparable to the decrease of federal fund rates we could argument for positive mean returns of the market. However, market uncertainty in times when bailouts as tool of last resort are necessary may not compare well to times of low economic growth when the central bank applies a standard tool to trigger growth. We could also hypothesize bank mean returns to respond most positively as bank liquidity is increased primarily, but we also have to bear in mind that government ownership in banks may lead to lower performance in the long run (see La Porta, Lopez-De-Silanes, and Shleifer, 2002).

Table 6.6 provides an overview on FSI and other industry average returns. We analyze the effect of the seven failure and the seven bailout dates on the industry subindices to increase statistical validity of our results. The S&P 500 Composite mean behavior is not displayed but shows negative mean returns around failure dates of which two are significant at the ten percent level. Respective mean returns around bailouts are not different from zero.

Financial services, industrials, and consumer industries are in line with the market observation showing even higher significance in their negative average returns around failure dates. Mean-wise financials show the strongest negative response to failures, but industrials median values are higher at times. Consumer goods and services are hardly less strongly affected. Also, for bailouts all four industries show basically zero mean returns with some single exceptions.

The findings only support parts of our expectations. Obviously, bank failures lead to decreasing asset values across industries but we cannot identify major differences and financial stocks do not clearly show strongest negative average returns. The scenario compares to that around bailouts where all industries and the market show basically zero returns. So bailouts can potentially stop the vicious circle of falling asset prices in the short run but they do not lead to positive returns on a market or an industry level.

Table 6.7 provides a view on four major financial services sub-sectors: REITs, banks, insurance companies, and asset managers.[18] Again, average returns around bank failures are homogeneously negative at high significance. However, around bailout dates banks realize some positive average returns the day before and on the bailout day itself. Positive average returns for the bailout date are also found for REITs and insurances, but they do show negative average returns in other event windows. Asset managers constitute the only financial subsector that continuously realizes negative average returns around bank bailouts.

[18] The fifth subsector considered is brokers but it is not displayed due to space constraints. It is the smallest sample comprised of only 19 single stocks. Brokers show consistently negative average returns around failures and insignificant average returns, statistically zero-returns, around bailouts.

Table 6.6: Average Returns of FSI and Other Industries around Bank Failures and Bailouts

Event	FSI (nobs=35)		IGI(nobs=49)		CGI (nobs=49)		CSI (nobs=28)	
wind.	Mean	Median	Mean	Median	Mean	Median	Mean	Median
Bank Failures								
0;0	-1.06% **	-0.62% *	-0.89% ***	-0.71% ***	-0.54% ***	-0.38% ***	-1.01% **	-0.10%
1;1	-0.38%	0.02%	-0.25% **	-0.16% *	-0.17%	-0.01%	-0.15%	0.16%
2;2	-0.67% ***	-0.35% ***	-0.54% **	-0.31% ***	-0.23% ***	-0.07% *	-0.44% ***	-0.02% **
1;-1	-0.40%	0.13%	-0.16%	0.25%	-0.18%	0.11%	-0.32%	0.50%
1;0	-0.73% **	-0.12% *	-0.52% ***	-0.49% ***	-0.36% ***	-0.34% ***	-0.66% ***	-0.30% **
2;0	-0.46% **	-0.19% *	-0.37% ***	-0.34% ***	-0.15% *	-0.03%	-0.48% ***	-0.18% ***
0;1	-0.37%	-0.17%	-0.30% ***	-0.04% *	-0.16%	-0.15% **	-0.07%	0.10%
0;2	-1.01% ***	-0.87% ***	-0.83% ***	-0.89% ***	-0.41% ***	-0.23% **	-0.59% ***	-0.39% ***
Bank Bailouts								
0;0	1.07%	-0.06%	0.95% *	-0.67%	1.03% *	0.04%	0.98%	-0.31%
1;1	-0.32%	-0.87%	-0.04%	-0.41%	0.07%	0.10%	-0.07%	-0.63%
2;2	-0.20%	0.02%	0.00%	-0.10%	0.06%	-0.05%	0.01%	-0.44%
1;-1	-0.75%	-0.86%	-0.73%	-0.74% *	-0.60%	-0.11%	-0.57%	-1.25%
1;0	0.16%	-0.57%	0.11%	0.21%	0.22%	0.04%	0.20%	-0.64%
2;0	0.13%	0.07%	0.05%	0.02%	0.16%	0.32%	0.03%	-0.02%
0;1	-0.11%	-0.01%	0.31%	-0.25%	0.40%	0.20%	0.18%	-0.38%
0;2	-0.11%	0.09%	0.28%	-0.13%	0.28%	0.10%	0.31%	-0.17%

Nobs: number of observations for each sample, failures and bailouts - calculated as number of indices multiplied each by seven failure and seven bailout dates; ***, **, and * indicate statistical significance at the 1%, 5%, and 10% levels, respectively; applied test statistics are the Wilcoxon signed rank test as non-parametric median test and the standard t-test as parametric mean test. FSI: Financial Services industry; IGI: Industrial Goods industry; CGI: Consumer Goods industry; CSI: Consumer Services industry.

Table 6.7: Average Returns of FSI Subsectors around Bank Failures and Bailouts

Event	REITs (nobs=833)		Banks (nobs=833)		Insurance (nobs=427)		Asset Mgrs. (nobs=210)	
wind.	Mean	Median	Mean	Median	Mean	Median	Mean	Median
Bank Failures								
0;0	-0.87% ***	-0.18% ***	-0.64% ***	-0.25% ***	-0.86% ***	-0.38% ***	-1.68% ***	-0.58% ***
1;1	-0.18% **	0.02%	-0.27% ***	-0.04% **	-0.53% ***	-0.14% ***	-0.96% ***	-0.45% ***
2;2	-0.54% ***	-0.22% ***	-0.67% ***	-0.29% ***	-0.79% ***	-0.20% ***	-1.19% ***	-0.46% ***
1;-1	-0.20%	0.17%	-0.12%	0.00%	-0.75% ***	-0.24% ***	-0.64% **	-0.30% **
1;0	-0.53% ***	-0.34% ***	-0.38% ***	-0.15% ***	-0.81% ***	-0.44% ***	-1.16% ***	-0.65% ***
2;0	-0.52% ***	-0.31% ***	-0.36% ***	-0.06% ***	-0.50% ***	-0.23% ***	-0.95% ***	-0.45% ***
0;1	-0.17% *	0.00% *	-0.35% ***	-0.07% ***	-0.42% ***	-0.07% ***	-1.12% ***	-0.52% ***
0;2	-0.66% ***	-0.40% ***	-0.96% ***	-0.55% ***	-1.10% ***	-0.37% ***	-1.59% ***	-0.92% ***
Bank Bailouts								
0;0	1.40% ***	0.09% ***	0.53% ***	0.00%	1.01% ***	0.64% ***	-0.20%	0.00%
1;1	-0.68% ***	-0.70% ***	0.05%	0.00%	-0.18%	-0.35% *	-0.65% ***	-0.49% ***
2;2	-0.43% ***	-0.16% ***	-0.03%	-0.08%	-0.11%	-0.11%	-0.43% **	-0.46% **
1;-1	-1.67% ***	-1.44% ***	0.21%	0.00%	-0.60% *	0.00% *	-0.32%	-0.01%
1;0	-0.13%	0.00%	0.37% ***	0.04% **	0.21%	0.03%	-0.26%	-0.06%
2;0	-0.16%	-0.04%	0.09%	0.00%	0.24% *	0.07%	-0.38%	-0.43% *
0;1	-0.18%	-0.14%	-0.03%	0.00%	0.03%	0.00%	-0.81% ***	-0.39% ***
0;2	-0.10%	-0.12%	0.04%	0.00%	-0.09%	-0.04%	-0.41%	-0.11%

Nobs: number of observations for each sample, failures and bailouts - calculated as number of stocks multiplied each by seven failure and seven bailout dates; ***, **, and * indicate statistical significance at the 1%, 5%, and 10% levels, respectively; applied test statistics are the Wilcoxon signed rank test as non-parametric median test and the standard t-test as parametric mean test.

Whereas on an industry level average returns where highly comparable across financials, industrials, and consumer goods and services around both, failures and bailouts, we find considerable asymmetries among the financial subsectors around bailouts. As hypothesized, we find evidence that only banks profit slightly form bank bailouts. Even though REITs share these positive average returns on the bailout day itself, they also show highly negative average returns in several other event windows. This indication of strong negativity is in line with evidence by Olliges, Raudszus, and Mueller (2011) who find negative abnormal REIT returns around bank bailouts. The fact that asset managers show consistently negative average returns around bailouts is remarkable as we did not formulate a respective expectation and find it rather hard to reason this observation.

6.4.2. Abnormal Stock Returns around Bank Events

Our second mean return analysis is an event study where we follow the objective to observe abnormal returns around the bank events. The estimation period of the event study comprises 250 trading days (one calendar year) and the regression keeps a distance to the actual event date of 30 trading days to minimize a potential bias through the event under analysis itself on the regression parameters. Event study methodology allows us to determine whether the mean return of some asset class significantly departs from its beforehand relationship to the market index (here S&P 500) around a certain event type.

We do not expect to observe considerable abnormal returns nor differences across industries around bank failures or bailouts as average returns of the market and all selected industries showed rather homogeneous behavior throughout the previous analysis. Within the financial service sector we do though expect to identify asymmetric abnormal returns. Similar to Raudszus, Olliges, and Mueller (2011) and Olliges, Raudszus, and Mueller (2011) we first expect REITs to again show positive abnormal returns around failures and negative abnormal returns around bailouts. Furthermore, Olliges, Raudszus, and Mueller (2011) observed negative abnormal returns also for other financial stocks. Because previous findings showed that only banks could realize positive average returns around bailouts we expect this finding to be mirrored in abnormal returns. This should primarily be opposed by asset managers as only they displayed negative average returns around bailouts. Regarding bank failures we most certainly assume to find negative abnormal returns for other (non-REIT) financial stocks as especially banks should face an increased risk of liquidity credibility.

In fact event study evidence on abnormal returns across industries is scarce and we decide not to display the respective table. We do not identify consistent abnormal returns that let us report any statistically solid findings on differences to zero or across industries. On the contrary, Table 6.8 reports again the asymmetric responses of FSI subsectors to bank failures and bank bailouts. Positive REIT abnormal returns around failures are determined as well as respective negative returns around bailouts. Especially insurance valuation contrasts REIT valuation as we find negative abnormal returns around failures and positive returns around bailouts. Banks show rather ambiguous mean shifts around both bank events with more significant results on the bailout part. Against expectations, asset managers hardly show any abnormal returns, which is the reason why we substituted them for brokers in the respective table. Brokers do show some consistent negative abnormal returns around bank bailouts.

Hence, previous evidence on REIT returns and bank events is confirmed. A closer look at banks' ambiguous behavior around bailouts uncovers that before the event abnormal returns are positive and from the bailout date onwards abnormal returns turn negative. This might indicate that some anticipation was not fully matched by the factual bailout and thus investors corrected their valuation. Insurance companies' unambiguous negative mean return reaction to failures and positive reaction to bailouts supports intuition and especially our expectations on failures. In sum, we cannot quite confirm pervious evidence by Olliges, Raudszus, and Mueller (2011) who found negative abnormal returns for REIT and non-REIT financial stocks around bailouts. In our analysis of single financial subsectors, extended by a day-before-the-event window, we uncover a rather asymmetric valuation picture.

Table 6.8: Abnormal Returns of FSI Subsectors around Bank Failures and Bailouts

Event	REITs (nobs=833)		Banks (nobs=833)		Insurance (nobs=427)		Brokers (nobs=133)	
wind.	Mean	Median	Mean	Median	Mean	Median	Mean	Median
Bank Failures								
0;0	0.37% ***	0.35% ***	0.22% *	0.24% ***	0.25%	0.33% ***	0.35%	0.49% **
1;1	0.33%	0.52% ***	-0.48% **	-0.29% *	-0.92% **	-0.51% ***	-0.60%	-0.58% *
2;2	0.88% ***	1.09% ***	-0.46%	0.11%	-1.09% **	-0.08%	-1.12%	-0.39%
1;-1	-0.08%	0.16%	-0.21%	-0.10%	-0.37% *	-0.31% ***	-0.01%	-0.01%
1;0	0.29% *	0.36% ***	0.01%	0.25%	-0.12%	0.17%	0.34%	0.23%
2;0	-0.22%	0.07%	-0.17%	0.20%	-0.49%	-0.08%	-0.07%	0.60%
0;1	0.40% **	0.35% ***	-0.28%	-0.17%	-0.56% *	-0.43% **	-0.59%	-0.49% **
0;2	1.46% ***	1.50% ***	-0.07%	0.33%	-0.35%	0.28%	-0.70%	-0.26%
Bank Bailouts								
0;0	0.47% **	-0.06%	-0.50% ***	-0.22% **	0.13%	0.25%	-0.18%	-0.33%
1;1	-1.22% ***	-1.69% ***	0.17%	0.45%	0.50%	0.25%	-0.51%	-0.73% *
2;2	-1.09% **	-1.49% ***	0.32%	0.70% *	0.31%	-0.38%	-1.37% **	-0.51%
1;-1	-0.75% ***	-0.56% ***	0.94% ***	0.49% ***	0.69% ***	0.48% ***	0.86%	0.40% **
1;0	-0.29%	-0.66% **	0.45% *	0.61% **	0.82% **	0.73% **	0.68%	0.10%
2;0	-0.11%	-0.47%	0.63% **	0.67% **	0.40%	-0.09%	0.50%	1.03%
0;1	-0.46%	-0.48% ***	-0.77% ***	-0.30% ***	-0.19%	-0.16%	-1.37% ***	-0.91% ***
0;2	-0.51%	-0.85% ***	-0.80% ***	-0.42% ***	0.04%	-0.02%	-2.05% ***	-1.64% ***

Nobs: number of observations for each sample, failures and bailouts - calculated as number of stocks multiplied each by seven failure and seven bailout dates; ***, **, and * indicate statistical significance at the 1%, 5%, and 10% levels, respectively; applied test statistics are the Wilcoxon signed rank test as non-parametric median test and the standard t-test as parametric mean test.

6.4.3. Stock Variance Adjustments around Bank Events

The third analysis step in general constitutes the first analysis of return risk and we start with an examination of adjustments in EGARCH variance estimates. As elucidated previously, daily index and stock variance estimates are determined applying the EGARCH (1,1) model. Consecutively, we determine daily changes in variances and cumulate them along the differently sized event windows. We perform this analysis for three major reasons: first, to scrutinize whether variance indeed decreases around bailouts and thus the government achieves its objected calming of the markets; second, we are eager to determine whether variance does react asymmetrically to the yet observed positive and negative average returns around bank events (leverage effect); and our third objective is to uncover any anticipation effect of a failure or bailout manifesting in a *calm-before-the-storm* effect.

To determine the hypothesized adjustment in estimated variances we compare our observed average returns from Table 6.6 and Table 6.7 to evidence by Ederington and Guan (2010) on the S&P 500 and how its EGARCH estimated volatility reacts to return shocks. They find that negative return shocks immediately and increasingly cause volatility to rise, but zero up to 2.5% positive returns do cause a decrease in EGARCH variance estimates. Positive return shocks of 3% and higher then again induce a rise in variance estimates which though is by far smaller than in the case of comparative negative return shocks.

As we derive our expectations from absolute (average) returns we again do not expect considerable differences across industries around bank failures and bank bailouts. Generally, we expect the significantly negative average returns around failures to be accompanied by significant variance increases. However, the day before the actual event, the event window [1;-1], does not show significance for any industry which implies basically zero returns and should thus result in a decrease of variance. We call this the *calm-before-the-storm* effect as the market anticipates a considerable negative announcement referenced to as *the storm*. In the case of bank bailouts we hardly observed any average returns different form zero which should cause variances to decrease across industries.

Table 6.9 shows considerable support to our expectations as we identify a series of positive variance shifts around failures across industries and FSI, CGI, and CSI truly show a significant *calm-before-the-storm* effect. In case of bank bailouts our expectations are also confirmed as we measure continuously negative variance shifts while the financial sector shows by far the strongest decrease as the primarily affected industry.

Table 6.9: Variance Adjustments of FSI and Other Industries around Bank Failures and Bailouts

Event	FSI (nobs=35)		IGI (nobs=49)		CGI (nobs=49)		CSI (nobs=28)	
wind.	Mean	Median	Mean	Median	Mean	Median	Mean	Median
Bank Failures								
0;0	.00000	-.00001	.00001	-.00001	.00000	.00000	.00001	-.00001
1;1	.00001	.00003	.00003 ***	.00003 ***	.00000	.00000	.00001	.00001
2;2	.00013 ***	.00012 ***	.00003 ***	.00002 ***	.00001	.00000	.00003 **	.00002 **
1;-1	-.00003	-.00003 ***	.00000	-.00001	.00000	-.00001 ***	.00000	-.00001 *
1;0	-.00003	-.00002	.00001	.00000	.00000	-.00001 **	.00001	-.00001
2;0	.00007 **	.00005 **	.00002 *	.00001 *	.00002 *	-.00001	.00003 **	.00000
0;1	.00004	.00002	.00003 *	.00002 **	.00001	.00000	.00002	.00001
0;2	.00006	.00003 *	.00001	.00002	.00000	.00000	.00001	.00000
Bank Bailouts								
0;0	.00004	-.00007	.00002	-.00003	.00002	-.00002	.00002	-.00001
1;1	-.00022 **	-.00021 **	-.00007 ***	-.00007 ***	-.00003	-.00005 **	-.00002	-.00005
2;2	-.00040 ***	-.00036 ***	-.00017 ***	-.00022 ***	-.00009 ***	-.00011 ***	-.00008 *	-.00011 **
1;-1	-.00016 ***	-.00017 ***	-.00003	-.00006	-.00002	-.00004 *	-.00001	-.00003
1;0	-.00012 *	-.00014 ***	.00000	-.00003	.00001	-.00003	.00001	-.00002
2;0	-.00015 *	-.00015 ***	-.00002	-.00004	-.00001	-.00004	.00000	-.00004
0;1	-.00006	-.00009 *	-.00005 **	-.00007 ***	-.00001	-.00005 **	-.00001	-.00005
0;2	-.00021 **	-.00016 **	-.00013 ***	-.00015 ***	-.00006 **	-.00008 ***	-.00006 *	-.00006 **

Nobs: number of observations for each sample, failures and bailouts - calculated as number of stocks multiplied each by seven failure and seven bailout dates; ***, **, and * indicate statistical significance at the 1%, 5%, and 10% levels, respectively; applied test statistics are the Wilcoxon signed rank test as non-parametric median test and the standard t-test as parametric mean test. FSI: Financial Services industry; IGI: Industrial Goods industry; CGI: Consumer Goods industry; CSI: Consumer Services industry.

The case of variance adjustments among the financial industry subsectors should be even more interesting. Whereas in the bank failure case we again do not observe differences in signs regarding average returns, there is considerable asymmetry around bailouts. Asset managers and REITs (except for the bailout day) show negative average returns, contrasting other financial stocks. Hence, we expect to observe positive variance adjustments for these asset classes in the respective event windows. Back to failures, REITs and banks are the only financial subsectors that show basically zero returns the day before bank insolvencies. Therefore, we should observe the *calm-before-the-storm* effect only for REITs and banks.

Table 6.10 provides the respective results.[19] There are several observations opposing our expectations: First, REITs show negative median variance shifts around bailouts, contrasting their respective negative average returns. Even though mean variance shifts appear partially positive, they are not significant. Second, also asset managers show negative median variance adjustments around bailouts while their respective returns are strictly negative. Again, positive mean variance shifts are not significant. Third, banks realize negative variance shifts around failures beyond the *calm-before-the-storm* effect. This contrasts their respective strictly negative average returns. And fourth, all financial subsectors experience a highly significant *calm-before-the-storm* effect, even though we only expected it for REITs and banks.

Our analysis on variance adjustments has again shown that generally the financial services industry does not behave much different than selective other industries except for its stronger decrease in variance around bailouts. However, the asymmetry among financial subsectors is prevalent and respective variance adjustments are much less predictive dependent on observed returns. Especially banks variance behavior seems to be affected also by other determinants than just the effective returns around bank failures. The fact that REITs and asset managers achieve variance descents while experiencing negative returns around bailouts might be caused by a spillover effect of other financial stocks such as banks and insurances where variance decreases strongly around bank bailouts.

---

[19] The fifth subsector, brokers, is not displayed due to space constraints. Brokers show strictly positive variance shifts around failures and strictly negative variance shifts around bailouts as would have been predicted from their respective average returns.

Table 6.10: Variance Adjustments of FSI Subsectors around Bank Failures and Bailouts

Event	REITs (nobs=833)		Banks (nobs=833)		Insurance (nobs=427)		Asset Mgrs. (nobs=210)	
wind.	Mean	Median	Mean	Median	Mean	Median	Mean	Median
Bank Failures								
0;0	-.00013	-.00001 ***	-.00003	-.00003 ***	.00005	-.00001	.00001	-.00002 ***
1;1	-.00015	.00005 ***	-.00003	-.00002 *	-.00022	.00000	.00023	.00001
2;2	.00008	.00011 ***	.00014 ***	.00005 ***	-.00030	.00006 ***	.00030	.00007 ***
1;-1	-.00004	-.00003 ***	-.00002	-.00003 ***	-.00034	-.00002 ***	.00003	-.00003 ***
1;0	-.00016	-.00001	-.00005	-.00003 ***	-.00029	-.00001	.00004	-.00002 *
2;0	.00007	.00005 ***	.00010 **	.00000 ***	-.00030	.00002 ***	.00004	.00001
0;1	-.00011	.00001 ***	-.00002	-.00002 ***	.00012	.00000	.00020	.00000
0;2	-.00011	.00002 ***	.00001	.00001 **	.00004	.00001 **	.00026	.00004 ***
Bank Bailouts								
0;0	.00015	-.00008	-.00002	-.00008 ***	-.00001	-.00004 *	-.00009	-.00004 **
1;1	-.00006	-.00020 ***	-.00016 ***	-.00011 ***	-.00045 **	-.00013 ***	.00018	-.00009 *
2;2	-.00033	-.00023 ***	-.00031 ***	-.00021 ***	-.00078 ***	-.00030 ***	-.00019	-.00014 ***
1;-1	.00017	-.00012 ***	-.00005	-.00007 ***	-.00024 ***	-.00009 ***	.00009	-.00014 ***
1;0	.00031	-.00011 ***	-.00007 *	-.00008 ***	-.00025 **	-.00009 ***	.00001	-.00009 ***
2;0	.00033	-.00020 ***	-.00006	-.00012 ***	-.00035 ***	-.00015 ***	-.00008	-.00009 **
0;1	-.00023	-.00017 ***	-.00011 ***	-.00010 ***	-.00021	-.00009 ***	.00008	-.00006 **
0;2	-.00051 *	-.00023 ***	-.00027 ***	-.00018 ***	-.00043 **	-.00015 ***	-.00019	-.00010 ***

Nobs: number of observations for each sample, failures and bailouts - calculated as number of stocks multiplied each by seven failure and seven bailout dates; ***, **, and * indicate statistical significance at the 1%, 5%, and 10% levels, respectively; applied test statistics are the Wilcoxon signed rank test as non-parametric median test and the standard t-test as parametric mean test.

6.4.4. Shifts in Asset-to-market Correlation around Bank Events

The fourth general and second risk analysis concentrates on adjustments in the correlation coefficient of the respective industries and financial subsectors to the S&P 500 as market index. We compute daily rolling correlation coefficients over 250 trading days (one calendar year). As, e.g., Bekaert and Wu (2000) have shown that correlation of firm returns tends to increase during downturns we are interested how correlation behaves around bank failures and bailouts as key-events of a financial crisis and related market downturn.

Due to a lack of previous evidence on asset-to-market correlation around bank failures and bailouts it is difficult to derive detailed hypotheses. We would expect that bank failures, which cause average returns across asset classes to be negative, drive market downturns and thus lead to increased correlation coefficients. As bailouts are surrounded by basically zero returns we assume a market downturn is decelerated at the moment and correlation coefficients stay rather constant.

However, as Table A6.4 shows across industries and Table 6.11 along FSI subsectors we find a rather uniform descent in correlation coefficients around bank failures and an increase in correlation coefficients around bank bailouts. FSI correlation coefficients though show a less strong reaction compared to IGI, CGI, and CSI around bailouts and basically no adjustments around failures. This is most certainly incurred by REIT correlation adjustments being positive in several event windows around failures which contrasts the strict negativity of other financial stocks. Another observation worth highlighting is the negativity in correlation shifts the day before bailouts for REITs and asset managers. Across industries this day shows strictly zero correlation shifts.

The analysis of adjustments in correlation coefficients has again shown that considerable asymmetry in reaction to bank events exists for financial stocks. Whereas banks, insurances, asset managers, and also brokers show a symmetric response in correlation shifts to bank failures and bailouts, REITs depart from that around failures. This fits well the fact that only REITs as financial stocks showed positive abnormal returns around failures, implying that investors interpret them as a *save haven*, resulting in a mean return shift in relation to the S&P 500. The negativity in REIT abnormal returns around bailouts is partially shared by banks and brokers which reasons that there are no discrepancies observable in respective correlation coefficient adjustments.

Table 6.11: Correlation Adjustments of FSI Subsectors around Bank Failures and Bailouts

Event	REITs (nobs=833)		Banks (nobs=833)		Insurance (nobs=427)		Asset Mgrs. (nobs=210)	
wind.	Mean	Median	Mean	Median	Mean	Median	Mean	Median
Bank Failures								
0;0	.0002	.0001 *	-.0009 ***	-.0001 ***	-.0026 ***	-.0006 ***	-.0012 **	-.0002 ***
1;1	.0003	-.0001	-.0008 *	-.0009 ***	-.0026 ***	-.0020 ***	.0005	-.0008
2;2	.0008	.0016 ***	-.0011 *	-.0013 ***	-.0020 **	-.0023 ***	.0000	-.0019 *
1;-1	-.0013 ***	-.0006 ***	-.0004 **	-.0002 *	-.0010 ***	-.0007 ***	-.0009 *	-.0004 ***
1;0	-.0011 ***	-.0003 ***	-.0014 ***	-.0005 ***	-.0036 ***	-.0020 ***	-.0021 ***	-.0011 ***
2;0	.0003	-.0002	-.0009 **	-.0007 ***	-.0021 ***	-.0017 ***	-.0016 *	-.0019 ***
0;1	.0016 ***	.0006 ***	-.0003	-.0005 **	-.0016 ***	-.0010 ***	.0015	.0003
0;2	.0007	-.0002	-.0011 **	-.0017 ***	-.0025 ***	-.0023 ***	.0005	-.0007
Bank Bailouts								
0;0	.0012 ***	.0010 ***	.0008 ***	.0006 ***	.0012 ***	.0010 ***	.0003	.0004 ***
1;1	.0018 ***	.0004 ***	.0017 ***	.0006 ***	.0018 ***	.0009 ***	-.0010	.0000
2;2	.0018 ***	.0018 ***	.0026 ***	.0021 ***	.0017 ***	.0013 ***	-.0018	.0013
1;-1	.0001	-.0003 **	.0005 ***	-.0001	.0002	-.0001	-.0009 **	-.0004 ***
1;0	.0013 ***	.0006 ***	.0013 ***	.0005 ***	.0014 ***	.0008 ***	-.0006	.0002
2;0	.0021 ***	.0015 ***	.0025 ***	.0017 ***	.0020 ***	.0016 ***	-.0009	.0007
0;1	.0017 ***	.0010 ***	.0012 ***	.0008 ***	.0016 ***	.0009 ***	.0000	.0004 ***
0;2	.0008 *	.0008 ***	.0009 ***	.0008 ***	.0010 **	.0006 ***	-.0006	.0006 **

Nobs: number of observations for each sample, failures and bailouts - calculated as number of stocks multiplied each by seven failure and seven bailout dates; ***, **, and * indicate statistical significance at the 1%, 5%, and 10% levels, respectively; applied test statistics are the Wilcoxon signed rank test as non-parametric median test and the standard t-test as parametric mean test.

6.4.5. Beta Shifts around Bank Events

The final univariate analysis concerns beta as systematic risk factor which we determine form variance and correlation estimates. Following the Capital Asset Pricing Model (CAPM) by Sharpe (1964) and Lintner (1965) beta is assumed to be stable throughout time. However, financial economists like Jagannathan and Wang (1996), Collins, Ledolter, and Rayburn (1987), Bos and Newbold (1984), Sunder (1980) and Fabozzi and Francis (1978) have shown that beta is conditional and follows a stochastic process. They thus argue against beta stability. We are eager to provide evidence on short-term beta changes around bank failures and bailouts to show that beta adjustments are significant and should not be neglected.

As Equation (6.2) shows, beta estimates are determined directly from daily variances and correlation coefficients which we have yet been analyzed and discussed. On the industry level we find that financial services experience no significant beta change. Variance shifts around failures are adverse and a variance descent around bailouts is compensated by the increase in correlation to the market. Interestingly, industrials' beta increases (variance driven) around failures as consumer goods' beta decreases (correlation driven). Consumer services' beta increases around bailouts (correlation driven). See Table A6.5 for details.

As Table 6.12 shows there is again asymmetry in beta adjustments across financial subsectors around failures. REITs' beta increase largely contrasts other financials' beta descent. Around bank bailouts we find a symmetric beta increase for all financials which though questions our previous finding on no significant beta changes for the FSI in general. Our results suggest that positive REIT abnormal returns around failures not necessarily lead to increased stock value as the beta risk increase may compensate the effect. Counterintuetively, banks may create value around failures as their basically unchanged mean return is accompanied by a significant beta risk descent in several event windows. Insurances' negative abnormal returns and negative risk shifts oppose REITs but should equally lead to a rather stable value. Scarce evidence on asset managers shows some positive abnormal returns and negative beta shifts enhancing its valuation. Brokers show rare evidence on abnormal returns and ambiguous beta risk adjustments implying a rather unchanged stock value.

REITs and brokers lose most value around bailouts as their beta risk increases add to negative abnormal returns. Asset managers with no abnormal returns and banks with ambiguous abnormal returns should experience a somehow smaller value descent. Only insurances with their positive abnormal returns around bailouts could potentially stabilize their value considering the significant beta risk increase.

Table 6.12: Beta Adjustments of FSI Subsectors around Bank Failures and Bailouts

Event	REITs (nobs=833)		Banks (nobs=833)		Insurance (nobs=427)		Asset Mgrs. (nobs=210)	
wind.	Mean	Median	Mean	Median	Mean	Median	Mean	Median
Bank Failures								
0;0	0.011 *	0.014 ***	-0.015 ***	0.002	0.011	0.004	-0.002	0.008 *
1;1	-0.007	0.009 *	-0.045 ***	-0.022 ***	-0.026	-0.032 ***	-0.020	-0.024 ***
2;2	0.009	0.020 ***	-0.030 ***	-0.007	-0.004	0.003	-0.018	-0.012
1;-1	0.013	0.011 ***	0.009 ***	0.006 ***	-0.004	0.006	0.004	0.007 *
1;0	0.025 *	0.029 ***	-0.006	0.004	0.008	0.008	0.002	0.008
2;0	0.026 *	0.029 ***	-0.016 **	0.010	0.013	0.014 **	-0.021	0.000
0;1	-0.020 **	-0.003 ***	-0.054 ***	-0.024 ***	-0.023	-0.033 ***	-0.024 **	-0.021 ***
0;2	-0.006	0.002	-0.030 ***	-0.007 ***	-0.006	-0.014 ***	0.001	-0.013
Bank Bailouts								
0;0	0.018 ***	0.019 ***	0.009 ***	0.002	0.011 *	0.010 **	0.005	0.000
1;1	0.025 *	0.034 ***	0.024 ***	0.013 ***	0.023 **	0.011 **	0.024	0.013 **
2;2	0.064 ***	0.032 ***	0.045 ***	0.019 ***	0.040 ***	0.021 ***	0.064 ***	0.032 ***
1;-1	-0.005	0.013 **	-0.002	0.001	-0.004	0.003	-0.004	0.002
1;0	0.013	0.011 *	0.007	-0.001	0.007	0.003	0.001	-0.001
2;0	0.061 ***	0.033 ***	0.035 ***	0.003 ***	0.032 ***	0.006 **	0.045 ***	0.006 **
0;1	0.030 ***	0.024 ***	0.026 ***	0.008 ***	0.027 ***	0.014 ***	0.028 **	0.015 ***
0;2	0.020 *	0.013 ***	0.018 ***	0.008 ***	0.019 *	0.008	0.025 *	0.020 ***

Nobs: number of observations for each sample, failures and bailouts - calculated as number of stocks multiplied each by seven failure and seven bailout dates; ***, **, and * indicate statistical significance at the 1%, 5%, and 10% levels, respectively; applied test statistics are the Wilcoxon signed rank test as non-parametric median test and the standard t-test as parametric mean test.

6.4.6. Determinants of Variance Adjustments and Daily Returns

Our sixth and final analysis consists of several multivariate regressions on daily returns and daily variance shifts of the financial services subindices versus the S&P 500 Composite. Hereby, we analyze the three daily returns and three daily variance shifts of the day before, the bank event day itself and the day after as dependent variables. Independent variables and potential determinants are *USD failure assets* and *USD bailout capital*, dummy variables, *D*, for the days *t-1*, *t*, *t+1* and the *day of the week* effect, the *S&P 500 daily return/variance shift*,[20] and the test for *autoregressive* returns and variances with FSI_{t-1} and $S\&P_{t-1}$ respectively. Equations (6.3) and (6.4) illustrate the models for daily returns, u_t, and daily variance shifts, $\Delta\sigma_t^2$:

$$u_t = \delta_1 + \delta_2 USD_{failure} + \delta_3 USD_{bailout} + \delta_4 D_{t-1} + \delta_5 D_t + \delta_6 D_{t+1} + \delta_7 \mathrm{D_{day}} \, [+ \, \delta_8 \mathrm{u}_{S\&P}] + \, \delta_9 \mathrm{u}_{t-1} \qquad (6.3)$$

$$\Delta\sigma_t^2 = \delta_1 + \delta_2 USD_{failure} + \delta_3 USD_{bailout} + \delta_4 D_{t-1} + \delta_5 D_t + \delta_6 D_{t+1} + \delta_7 D_{day} \, [+ \, \delta_8 \Delta\sigma_{S\&P}^2] + \, \delta_9 \Delta\sigma_{t-1}^2 \qquad (6.4)$$

Our first analysis has shown that daily FSI and S&P 500 returns respond negatively to failures and insignificantly to bailouts. We expect to confirm these observations. Return-wise we have not observed equivalent average returns for the three days around failures and bailouts. Thus, we do not expect to observe significances here.

Our third analysis has shown evidence for generally positive variance shifts around failures but also a significant *calm-before-the-storm* effect. Overall, bailouts incurred decreasing variances. Thus, we expect to confirm these observations on δ_2 and δ_3. Also, we may expect a negative δ_4 because variance strictly decreases the day before failures and bailouts. However, we do not expect to observe a general bank event effect nor a significant general day-after effect.

Table 6.13 provides the results for our multivariate regressions on financial services and S&P 500 returns around bank failures and bank bailouts. We performed three regression models each to test all dependent variables avoiding autocorrelation. Models 2 and 3 and models 5 and 6 show that FSI and S&P 500 returns respond negatively to failures and positively to bailouts. Our expecta-

---

[20] Only applicable when analyzing the financial services returns and variance shifts.

tions on δ_2 and δ_3 are therefore as much confirmed as by the insignificances of δ_5 and δ_6. The multivariate regressions do not find significance for δ_4 which supports evidence from our first analysis on insignificant day-before-the-event returns. Furthermore, financials' returns show strong dependence on S&P returns, the S&P 500 return is affected negatively by the weekday Monday around bank events and it appears negatively autoregressive at low significance.

Table 6.13: Multivariate Regressions on FSI and the S&P 500 Returns around Bank Events

		Financial Services Industry			S&P 500 Composite		
Coefficients		Model 1	Model 2	Model 3	Model 4	Model 5	Model 6
Intercept	δ_1	(0.00)	(0.01) ***	(0.00) *	0.01	(0.01) *	(0.01)
USD failure assets	δ_2		(0.00) ***	(0.00) ***		(0.00) ***	(0.00) ***
USD bailout capital	δ_3		0.00 ***	0.00 ***		0.00 ***	0.00 ***
t-1	δ_4		0.01			0.00	
t	δ_5	0.00			(0.00)		
t+1	δ_6			(0.00)			0.00
Monday	δ_7				(0.02) ***		
S&P500	δ_8	1.31 ***					
FSI_{t-1}, $S\&P_{t-1}$	δ_9	0.03	(0.05)	(0.05)	(0.27) *	(0.05)	(0.06)
Nobs.		205	205	205	41	41	41
F		0.00 ***	0.00 ***	0.00 ***	0.11	0.00 ***	0.00 ***
adj. R^2		0.73	0.24	0.24	0.11	0.40	0.40
DW		2.00	1.90	1.90	2.02	1.88	1.88

(x) values in parentheses are negative, Nobs: number of observations - three trading days (*t-1, t, t+1*) per event (seven failures and seven bailouts) and index (five subindices for financial services).

In case of FSI and S&P daily variance shifts as dependent variable in multivariate regressions, supporting evidence for our previous analyses is less easily identified. Table A6.6 shows insignificant FSI variance responses to failures which is in line though with the ambiguous respective evidence from our previous analysis. In case of FSI variance and bailouts a significantly negative δ_2 could have been expected, however, variance descents of REITs and asset managers were not that reliable. Thus, also an insignificant δ_2 does not contrast previous findings. Only in the case of δ_4 our expectation to observe a significant *calm-before-the-storm* effect has not been met. In addition, model 5 regarding S&P 500 variance finds a negative effect through failures which does not really fit ambiguous previous evidence. Else, we observe a negative *mid-of-the-week* effect (Wednesday effect) on FSI variance shifts, strong positive dependence on S&P 500 variance shifts and significant negative autoregressiveness. Also the S&P 500 variance shifts display significant negative autoregressiveness.

6.5. Conclusion

Our study has the objective to show how the financial services industry's (FSI) return performance responds to bank bailouts of the recent financial crisis controlling for bank failures. We analyze stock return performance by scrutinizing shifts in mean returns and adjustments in the three risk measures variance, correlation to the market, and beta. For robustness, we compare FSI performance to the performance of the general market (S&P 500) and three exemplary other industries: industrial goods, consumer goods, and consumer services. Core of our examination though constitutes the FSI and its subsectors REITs, banks, insurances, asset managers, and brokers.

We contribute to the literature in three major parts: First, we find evidence that government bailouts indeed achieve a short-term decrease in market volatility and, we do not find evidence that bailouts distribute exceptional wealth to holders of financial stock. However, decreased volatility is paralleled by increasing beta as systematic risk measure for all FSI subsectors which supports moral hazard theory. Second, we show that FSI subsectors react asymmetrically to bank failures and bank bailouts in mean returns and return risk. Especially REITs depart from other financial stock performance various times. And third, the application of the EGARCH (1,1) model to estimate variance lets us show that variance indeed reacts asymmetrically to positive and negative return shocks and proves the existence of the so called leverage effect.

Our finding of generally increasing beta measures around bank bailouts is a strong sign for moral hazard. Even though we identified strong and highly significant variance descents across all financial subsectors, a comparably uniform increase in correlation to the market (S&P 500) must have offset the former effect. This finding extends previous evidence by Gropp, Hakenes, and Schnabel (2011) who found a strong decrease in risk adversity of competitor banks in response to bank bailouts and guarantees.

The asymmetry in reactions to bank events by financial stocks is primarily caused by the behavior of REITs. E.g., REITs positive abnormal returns around failures contrast its peers. In this case its behavior supports the *save haven-theory* of Mueller and Mueller (2003) that sees real estate as a hard asset and thus preferred investment option during times of high market uncertainty. However, REITs also depart from the strictly negative correlation shift of financial stocks around failures which then leads to a sole increase in beta risk around failures. Consequently, we cannot say whether REITs' two-sided shift in risk-return profile finally leads to a value increase or decrease around bank failures.

Finally, the application of the EGARCH model has proven to be appropriate in describing stock market variance in our study. We found that the asymmetry in volatility response to return shocks, a stylized fact of stock return data, describes well the behavior of our financial stocks' variances. We found strong variance increases around the negative (average) return shocks induced by bank failures and significant variance descents around zero and small positive returns induced by bank bailouts. This adds well to recent evidence by Ederington and Guan (2010) modeling S&P 500 variance with an EGARCH model. Furthermore, we could identify a significant *calm-before-the-storm* effect implying that markets anticipate bank failures which adds well to evidence by Bomfim (2003) who shows comparable evidence on the anticipation of federal fund rate adjustments.

As a result of our paper we recommend policy makers to be concerned about the backwater they create bailing out banks to calm down financial markets on a short-term basis. Increasing asset-to-market correlation offsets short-term descents in variances and leads to increased systematic risk of financial services subsectors and other industries. This increased level of risk looms over the economy as a whole and can propagate in the aftermath of a financial crisis which is often characterized by a government debt crisis (see Reinhart and Rogoff, 2009).

We like to point out three recommendations for further research that constitute limitations of the present study. First, future research could examine the general explanatory power of our findings by analyzing another set of bank failures and bailouts of either an earlier financial crisis or with a distinct geographic focus. Second, financial economists may want to extend the view from the financial services industry to other industries of the general market and their corresponding subsectors. And third, researchers could advance our methodological approach by, e.g., limiting their analysis to variance shifts applying a diverse set of GARCH models and competing approaches to validate consistency of findings and to enhance the general robustness of results.

7. Concluding Remarks

My studies in the preceding chapters have shown that shareholder wealth is strongly affected by management decision making and external events. I find considerable adjustments in both, first and second moments of mean returns with respect to M&A announcements and banking crisis events. Yet previous evidence has shown that the mean return as first moment of returns may experience significant shifts with respect to management decisions and external market events. There is also earlier evidence on, e.g., beta risk dynamics dependent on comparable events. However, I close a research gap in that I scrutinize both simultaneously concentrating on short-term effects. As the theory of efficient markets says new information is immediately reflected in actual stock prices, I regard the short-term behavior of stocks and indices as the key indicator on how capital markets evaluate the new status quo.

The event study methodology has since decades been a prominent tool among scholars of multiple disciplines to measure an event's effect on the first moment of stock returns: the mean return. The respective model, the market model, builds on the basic assumption that the beta factor, also denoted as an asset's systematic risk indicator and thus, determinant of the asset's expected return relative to the general market, is stable throughout time. Many researchers have though argumented against beta stability and shown that beta behavior is actually stochastic and therefore conditional. I model beta behavior on the basis of time-variant asset and market volatility estimates and the prevalent correlation among these two. The so called GARCH models incorporate certain stylized facts of stock market data. First among these is the characteristic of volatility clustering, meaning that there is a strong autoregressive component in volatility. Second, the further advanced EGARCH model allows to additionally control for the asymmetry in volatility reaction to positive and negative return shocks as also this is a proven stylized fact of stock market data.

The dissertation constitutes an illustrative development in methodology from standard event study and risk measurement techniques of previous evidence to the application of a fully integrated return and risk event analysis applying the EGARCH model:

In my first study, reexamining on the issue of corporate diversification, I first perform a standard event study to confirm previous findings building on an adjusted data sample. I advance the shareholder wealth analysis by additionally addressing beta risk dynamics in the short and long run. The long-term beta risk analysis is comparable to previous research as it applies standard OLS regressions to obtain beta estimates and respective adjustments over several one year intervals. I mirror this analysis' results applying the GARCH model. The true advantage of the GARCH model though constitutes the fact that it allows to determine daily variance and respective beta measures without limiting the estimation period to a (randomly) defined time frame. This methodological characteristic enables the scholar to measure short-term risk shifts around an event under examination. Hence, I can analyze the second moment of stock returns in a comparable format as the first moment of stock returns is scrutinized by the standard event study.

My second study again investigates M&A announcements but this time limited to horizontal transactions with a special focus on a strongly consolidating industry and corresponding rival effects. As a respective previous study has yet analyzed first moment of stock returns, I concentrate on beta risk as part of second moment of stock returns on the basis of a highly comparable data sample. The application of the GARCH model again allows to observe short-term adjustments in systematic risk for acquirers and rivals and adds well to the literature.

The studies three to five shift focus to external events and their potential effect on shareholder wealth of certain corporations. The great financial crisis beginning in the year 2007 made me choose bank failures and bank bailouts as core events of this crisis. I expected significant changes in shareholder wealth regarding certain financial stocks as these are highly dependent on market liquidity which's availability and price shows dependency on the named bank events. In chapters four and five I investigate bank failures' and bailouts' implications on short-term first and second moment of REIT stock returns.

Finally, my last study discusses the adverse effects bank bailouts versus failures have shown on REITs on the financial services industry in general. Hereby, I extend the analyses on first and second moment of returns and advance the volatility modeling by applying the EGARCH method. In fact, I examine short-term behavior of plain mean returns and abnormal returns so as variance rates, correlation coefficients, and the respective beta factors. I find clear evidence that the sample stocks and indices show the expected asymmetry in volatility reaction to comparable positive and negative return shocks as a typical stylized fact of financial stock return data.

Apart from the methodological perspective of this dissertation, and most probably of higher interest to many readers of this book and the studies it comprises, remain the actual adjustments in shareholder wealth I uncover:

My first study on corporate diversification strategies in the construction industry builds on previous findings which indicate that vertical diversification is beneficial to lateral based on event study analysis. I am able to confirm these first moment of stock return findings for the sample noting a generally positive bias induced by the economic period 2002-2007. Furthermore, the risk event study finds that lateral M&A actually induce a beta risk increase in the short run around the transaction announcement. Vertical M&A show to be beneficial to acquirers in the long run. I uncover a significant asset beta descent from the year prior to two years post M&A announcements. Consequently, my findings on beta behavior, as part of second moment returns, do not falsify previous evidence but underline the superiority of vertical diversification in the construction industry.

My second study again examines M&A as a strategic management decision and its effect on shareholder wealth. Here, I investigate only horizontal transaction announcements as I am curious to learn how in a strongly consolidating industry the announcement of a takeover, under the constraint of a limited number of potential mature and emerging market targets, affects acquirer and rival systematic risk. A previous analysis on the same data set yet finds positive acquirer abnormal returns and negative rival abnormal returns. My findings on systematic risk dynamics again add well in that acquirers experience a decline and missing-out rivals an increase in beta. Acquirer risk declines are exceptionally high in cross-border M&As with targets in emerging markets, whereas rivals suffer most missing out on one of the scarce opportunities in mature markets. Again, the short-term risk analysis proves to be highly valuable in measuring shareholder wealth effects and systematic risk factor beta should not simply be assumed constant.

My third study examines the effect of bank failures as external events on REITs. On the one hand REITs are considered as financial stock and are heavily dependent on market liquidity. On the other hand they are seen as good estimators for common equity behavior due to their high dividend payout ratio unable to apply return biasing dividend strategies. As previous evidence around monetary shocks has found their behavior comparable to other equities, I investigate the case when the shock is unexpected, from the inside of the system, and can potentially decrease liquidity supply strongly, e.g., the case of bank failures. In fact, I find evidence that especially Equity REITs experience positive abnormal returns relative to the market, while risk levels are basically stable. This supports the save haven theory because Equity REITs are seen as real estate and hard asset during times of crisis which benefits their valuation relative to common equities.

My fourth study investigates the other side of the medal in that it scrutinizes the effect of bank bailouts on REITs. Bank bailouts are also an external event, more or less expected by market participants, and their impact on market liquidity should affect REIT shareholders' wealth. However, bank bailouts are always accompanied by a discussion on moral hazard as they are seen to provoke excessive risk taking. Bad speculation may not necessarily cause failure of the respective institute but can in fact lead to higher returns if a bailout takes place or can be expected. Hence, REITs as hard assets could either be outperformed by more risky general equities or benefit as market liquidity is substantially increased and their need for external financing may be better satisfiable. I find REITs underperform the market on the return side, while systematic risk increases slightly. Other financial stocks show a comparable behavior. The results are driven by Equity REITs that carry most hard asset real estate characteristics and thus clearly support the save haven theory.

My fifth and final study, discussed in chapter six of this dissertation book, brings together the approaches of the two previous studies and advances the risk modeling. As the previous studies have shown opposing mean return effects of bank failures and bank bailouts on REITs, I am curious whether respective short-term variance adjustments are truly asymmetric. The other studies of this dissertation focus on beta risk as part of second moment returns as it is the most common risk estimator. Beta relates directly to the discussion on shareholder wealth as it is a variable of the standard market model and therefore a pricing factor for expected equity returns. My analyses show that standard mean returns around failures are negative and do not deviate across financial stocks (e.g., REITs, banks, and insurances). However, the fact that REITs experience negative mean returns also around bailouts opposes other financial stocks. These findings translate to the yet seen positive abnormal REIT returns around failures and negative abnormal returns around bailouts differentiating significantly from other financial stocks. EGARCH estimated variances largely show the expected asymmetry in that negative average returns lead to higher variances, whereas zero and smaller positive mean returns even incur a variance decline. Contrary to intuition, beta still rises across financial stocks around bailouts driven by the steep increases in asset-to-market correlation that offsets partial variance descents. Hence, moral hazard theory finds support in the data analysis as the potential decrease in short-term volatility still comes at the price of increased systematic risk.

Overall, this dissertation shows that risk dynamics are quite significant even in the short run. Thus, evaluating the effect of any event on shareholder wealth scholars should not limit their view and analysis scope to shifts in mean returns as first moment of stock returns. There is a wide set of methodological approaches to measure risk dynamics and the use of models from the GARCH family is not compulsory. However, I am convinced that it is one appropriate approach to estimate and observe short-term adjustments in risk measures such as variance for total risk and beta for systematic risk.

I draw three major implications from this dissertation project. These are, in my opinion, relevant to corporate managers and corporate and private investors. I relate the implications to quotes of outstandingly successful business men who have proven their superior business sense and are generally seen as among the greatest achievers of their fields.

"Wide diversification is only required
when investors do not understand what they are doing."
Warren Buffet

Portfolio theory says that diversification is best achieved at the level of the investor holding a mix of the market portfolio and risk free assets. This allows the best risk-return profile adjusted for a persons' individual risk adversity characteristic without the necessity of actively managing funds. In theory, this describes probably the optimal investment strategy for most private investors that do not possess superior understanding of market dynamics and who's area of expertise lies somewhere else.

Corporate managers, however, should have gained exceptional expertise on their company, the competitive landscape, and their industry in general. To strengthen their company's competitive advantage they must invest on the basis of their understanding which though presumably decreases the further potential opportunities are away from the focus of their expertise. Hence, corporate managers must weigh potential benefits from diversification controlling for their limited knowledge.

My first study shows that for the construction industry the line of value creation versus value destruction lies between vertical and lateral diversification. Even though greater diversification potentially decreases operative business risk, capital market participants do not trust that such a wide stretch of expertise will be successful and discount the respective strategic option. I interpret this as evidence that diversification has its clear limits at the corporate level.

"A competitive world offers two possibilities.
You can lose. Or, if you want to win, you change."
Henry Ford

Being successful when the economy as whole and the respective market a corporation finds itself in grow, does not constitute a major challenge. Corporate managers may achieve business growth by administering existent client relationships and targeting for the respective share of new clients in the growing market.

Unfortunately, few markets may be characterized by such a dynamic and unless entry barriers are prohibitively high competition increases rapidly, resulting in a strong pressure on costs as prices are decreased by a product oversupply. The solar industry is probably a good example for the latter case. On the contrary, the brewing industry faces several markets with declining demand, foremost in mature economies. My second study shows that in a contracting market competition becomes exceptionally fierce as external growth appears yet as a final option to offset the market decline and achieve by minimum a stable profitability.

The evidence shows that the rivalry, especially among the four international brewery conglomerates, is strong as the potential for horizontal acquisitions declines. The survivorship probability of those who miss out on the recent opportunities is clearly downgraded by capital markets as their risk-return profile suffers on both sides. So corporate managers might as well consider a change or adjustment of their business approach to raise the amount of growth opportunities and achieve a positive corporate outlook.

"Real Estate is at the core of almost every business,
and it's certainly at the core of most people's wealth.
In order to build your wealth and improve your business smarts,
you need to know about real estate."
Donald Trump

Real Estate Investment Trusts (REITs) constitute a special legal entity type that may be listed at the stock exchanges. Its name yet indicates its concentration on real estate investments and thus, its special status among listed companies. There is an extensive discussion among scholars of real estate economics whether REITs do behave more like common stocks or rather like its fundamental investment being real estate. Furthermore, times of high market volatility have shown to significantly impact REITs' behavioral characteristics.

To corporate investors as to private investors REITs appear as an attractive opportunity to invest into real estate linked investment facilities achieving an increased diversification across properties compared to investing in a single real estate project and ensuring a high liquidity of the investment comparable to common equities. The not fully informed investor may though assume that REITs are basically real estate and thus interpret it as a hard asset, serving as inflation hedge in turbulent times representing a save haven for surplus liquidity.

My analyses on REITs, classified as financial stocks, and their respondence to the banking crisis events bank failures and bailouts show that there are considerable differences among the various REIT types. I find that foremost Equity REITs and within this category residential and healthcare focused REIT types show most hard asset and save haven characteristics when market uncertainty increases around bank failures. In contrast, Mortgage REITs that invest in mortgage backed loans and do not possess real properties behave more like other equities. Generally, I find that REITs may also show higher volatility than other common equities. The implication for investors is to precisely scrutinize REIT investments before assuming plain proximity to real estate valuations.

Henceforth, I like to elaborate on two strains of recommendations for further research adding to this set of studies: First, scholars of financial economics or related disciplines may find it valuable to invest further research efforts into the advancement of the applied methodology in this dissertation. On the one hand this concerns the event study. Its appropriateness depends largely on the sample selection and the sample size. I yet applied bootstrapping to the event study results in the first study as this was recommended by an anonymous referee in the paper submission process due to the limited size of the sample. On the other hand researchers could measure the relative performance of some of the many other GARCH models the GARCH family comprises with respect to the measurement of systematic risk. This approach could be further extended by broadening the scope to beta risk modeling techniques outside from the GARCH family.

Second, future research could build upon and extend the findings of my studies. Potential research questions may concern corporate diversification and wealth effects in different industries or even across industrial sectors; risk dynamics in the globally consolidating automotive or logistics industry, where target companies are scarce and potential cost and sales synergies are high; REIT behavior and affiliation to common equities, financial stocks, or an individual asset class in the face of crisis types such as, e.g., terrorist attacks or natural catastrophes. Finally, for this set of commendations, there is much research left that actually investigates the asymmetric volatility behavior of single stocks and certain industries. This is most definitely the area of research where I see myself when I get the chance and decide to expand my academic career.

References

Acharya, Viral, Irvind Gujral, Nirupama Kulkarni, Hyun Song Shin, 2011, Dividends and Bank Capital in the Financial Crisis of 2007-2009, NBER working paper No. 16896.

Acharya, Viral, Thomas Philippon, Matthew Richardson, Nouriel Roubini, 2009, The Financial Crisis of 2007-2009: Causes and Remedies, Financial Markets, Institutions and Instruments 18/2, 89-137.

Adams, William J., 2006, Markets: Beer in Germany and the United States, Journal of Economic Perspectives 20/1, 189-205.

Adrian, Tobias, Joshua Rosenberg, 2008, Stock Returns and Volatility: Pricing the Short-Run and Long-Run Components of Market Risk, Journal of Finance 63/6, 2997-3030.

Aït-Sahalia, Yacine, Robert Kimmel, 2007, Maximum likelihood estimation of stochastic volatility models, Journal of Financial Economics 83, 413-452.

Ambrose, Brent W., Dong Wook Lee, Joe Peek, 2007, Comovement After Joining an Index: Spillovers of Nonfundamental Effects, Journal of Real Estate Economics 35/1, 57-90.

Amihud, Yakov, Baruch Lev, 1981, Risk reduction as a managerial motive for conglomerate mergers, Bell Journal of Economics 12/2, 605-617.

Amihud, Yakov, Gayle L. DeLong, Anthony Saunders, 2002, The effects of cross-border bank mergers on bank risk and value, Journal of International Money and Finance 21/6, 857-877.

Amit, Raphael, Joshua Livnat, 1988a, Diversification strategies, business cycles, and economic performance, Strategic Management Journal 9/2, 99-110.

Amit, Raphael, Joshua Livnat, 1988b, Diversification and the risk-return tradeoff, Academy of Management Journal 31/1, 154-166.

Amit, Raphael, Birger Wernerfelt, 1990, Why do firms reduce business risk?, Academy of Management Journal 33/3, 520-533.

Ball, Ray, Philip Brown, 1968, An Empirical Evaluation of Accounting Income Numbers, Journal of Accounting Research 6/2, 159-178.

Banal-Estañol, Albert, Marco Ottaviani, 2006, Mergers with Product Market Risk, Journal of Economics and Management Strategy 15/3, 577-608.

Banal-Estañol, Albert, Jo Seldeslachts, 2011, Merger Failures, Journal of Economics and Management Strategy 20/2, 589-624.

Basse, Tobias, Meik Friedrich, 2009, REITs and the Financial Crisis: Empirical Evidence from the U.S., International Journal of Business and Management 4/11, 3-10.

Beitel, Patrick, Dirk Schiereck, Mark Wahrenburg, 2004, Explaining M&A Success in European Banks, European Financial Management 10/1, 109-139.

Bekaert, Geert, Guojun Wu, 2000, Asymmetric volatility and risk in equity markets, Review of Financial Studies 13/1, 1-42.

Bernanke, Ben S., Kenneth N. Kuttner, 2005, What Explains the Stock Market's Reaction to Federal Reserve Policy?, Journal of Finance 60/3, 1221-1257.

Boehmer, Ekkehart, Jim Musumeci, Annette B. Poulsen, 1991, Event-study methodology under conditions of event-induced variance, Journal of Financial Economics 30/2, 253-272.

Bollerslev, Tim, 1986, Generalized autoregressive conditional heteroscedasticity, Journal of Econometrics 31/3, 307-327.

Bollerslev, Tim, 1990, Modelling the Coherence in Short-Run Nominal Exchange Rates: A Multivariate Generalized Arch Model, Review of Economics and Statistics 72/3, 498-505.

Bomfim, Antulio N., 2003, Pre-announcement effects, news effects, and volatility: Monetary policy and the stock market, Journal of Banking and Finance 27/1, 133-151.

Bos, Theodore, Paul Newbold, 1984, An Empirical Investigation of the Possibility of Stochastic Systematic Risk in the Market Model, Journal of Business 57/1, 35-41.

Bredin, Don, Gerard O'Reilly, Simon Stevenson, 2007, Monetary Shocks and REIT Returns, Journal of Real Estate Finance and Economics 35/3, 315-331.

Brooks, Robert D., Robert W. Faff, Michael D. McKenzie, 1998, Time-Varying Beta Risk of Australian Industry Portfolios: A Comparison of Modeling Techniques, Australian Journal of Management 23/1, 1-22

Calomiris, Charles W., 1995, Financial Fragility: Issues and Policy Implications, Journal of Financial Services Research 9/3-4, 241-257.

Carlson, Mark A., Thomas King, Kurt Lewis, 2011, Distress in the Financial Sector and Economic Activity, B.E. Journal of Economic Analysis and Policy 11/1, 1-29.

Cecchetti, Stephen G., 2009, Crisis and Responses: The Federal Reserve in the Early Stages of the Financial Crisis, Journal of Economic Perspectives 23/1, 51-75.

Chatterjee, Sayan, Michael Lubatkin, 1990, Corporate mergers, stockholder diversification, and changes in systematic risk, Strategic Management Journal 11/4, 255-268.

Chen, YiTing, Chung-Ming Kuan, 2002, Time Irreversibility and EGARCH Effects in US Stock Index Returns, Journal of Applied Econometrics 17/5, 565-578.

Choi, Jongsoo, Jeffrey Russel, 2004, Economic gains around mergers and acquisitions in the construction industry of the United States of America, Canadian Journal of Civil Engineering 31/3, 513-525.

Clayton, Jim, Greg MacKinnon, 2001, The Time-Varying Nature of the Link Between REIT, Real Estate and Financial Asset Returns, Journal of Real Estate Portfolio Management 7/1, 43-54.

Clougherty, Joseph A., Tomaso Duso, 2009, The Impact of Horizontal Mergers on Rivals: Gains to Being Left Outside a Merger, Journal of Management Studies 46/8, 1365-1395.

Collins, Daniel W., Johannes Ledolter, Judy Rayburn, 1987, Some Further Evidence on the Stochastic Properties of Systematic Risk, Journal of Business 60/3, 425-448.

Conn, Robert L., 1985, A Re-examination of Merger Studies That Use the Capital Asset Pricing Model Methodology, Cambridge Journal of Economics 9/1, 43-56.

Connell, Fred, Robert L. Conn, 1993, A Preliminary Analysis of Shifts in Market Model Regression Parameters in International Mergers Between US and British Firms: 1970-1980, Managerial Finance 19/1, 47-77.

Cordella, Tito, Eduardo L. Yeyati, 2003, Bank bailouts: moral hazard vs. value effect, Journal of Financial Intermediation 12/4, 300-330.

Corrado, Charles J., 2011, Event studies: A methodology review, Accounting and Finance 51/4, 207-234.

Corrado, Charles J., Cameron Truong, 2008, Conducting event studies with Asia-Pacific security market data, Pacific-Basin Finance Journal 16/5, 493-521.

Cuervo, Javier C., Low Sui Pheng, 2005, Significance of internationalization factors for Singapore transnational construction corporations, Construction Management and Economics 23/2, 147-162.

Darrat, Ali F., Shafiqur Rahman, Maosen Zhong, 2003, Intraday trading volume and return volatility of the DJIA stocks: A note, Journal of Banking and Finance 27/10, 2035-2043.

Davidson, III, Wallace N., Sharon H. Garrison, Glenn V. Henderson Jr., 1987, Examining Merger Synergy with the Capital Asset Pricing Model, Financial Review 22/2, 233-247.

Davis, Rachel, Irene M. Duhaime, 1992, Diversification, industry analysis and vertical integration: New perspectives and measurement, Strategic Management Journal 13/7, 511-524.

Delany, Frank T., Sam C. Wamuziri, 2004, The impact of mergers and acquisitions on shareholder wealth in the UK construction industry, Engineering, Construction and Architectural Management 11/1, 65-73.

Dell'Ariccia, Giovanni, Enrica Detragiache, Raghuram Rajan, 2008, The real effect of banking crises, Journal of Financial Intermediation 17/1, 89-112.

Demyanyk, Yuliya, Otto V. Hemert, 2011, Understanding the Subprime Mortgage Crisis, Review of Financial Studies 24/6, 1848-1880.

Diamond, Douglas W., Raghuram Rajan, 2005, Liquidity Shortages and Banking Crises, Journal of Finance 60/2, 615-647.

Dodd, Peter, 1980, Merger proposals, management discretion and stockholder wealth, Journal of Financial Economics 8/2, 105-137.

Dos Santos, Marcelo B., Vihang R. Errunza, Darius P. Miller, 2008, Does corporate international diversification destroy value? Evidence from cross-border mergers and acquisitions, Journal of Banking and Finance 32/12, 2716-2724.

Earlam, Melissa, Jason DeRise, Renier Swanepoel, Kaumil Gajrawala and Poelma Zachary, 2010, European Beverages & Breweries - 2010: A flight to quality (Equity Research Report), London.

Ebneth, Oliver, Ludwig Theuvsen, 2007, Large mergers and acquisitions of European brewing groups – event study evidence on value creation, Agribusiness 23/3, 377-406.

Eckbo, B. Espen, 1983, Horizontal Mergers, Collusion, and Stockholder Wealth, Journal of Financial Economics 11/1-4, 241-273.

Eckbo, B. Espen, 1985, Mergers and the Market Concentration Doctrine: Evidence from the Capital Market, Journal of Business 58/3, 325-349.

Ederington, Louis H., Wei Guan, 2010, How asymmetric is U.S. stock market volatility?, Journal of Financial Markets 13/2, 225-248.

Elgers, Pieter T., John J. Clark, 1980, Merger Types and Shareholder Returns: Additional Evidence, Financial Management 9/2, 66-72.

Ellis, Luci, 2010, The Housing Meltdown: Why Did It Happen in the United States?, International Real Estate Review 13/3, 351-394.

Elsas, Ralf, Andreas Hackethal, Markus Holzhäuser, 2010, The anatomy of bank diversification, Journal of Banking and Finance 34/6, 1274-1287.

Engle, Robert F., 1982, Autoregressive conditional heteroscedasticity with estimates of the variance of UK inflation, Econometrica 50/4, 987-1008.

Engle, Robert F., Victor K. Ng, Measuring and Testing the Impact of News on Volatility, 1993, Journal of Finance 48/5, 1749-1778.

Fabozzi, Frank J., Jack C. Francis, 1978, Beta as a random coefficient, Journal of Financial and Quantitative Analysis 13/01, 101-116.

Faff, Robert W., David Hillier, Joseph Hillier, 2000, Time Varying Beta Risk: An Analysis of Alternative Modelling Techniques, Journal of Business Finance and Accounting 27/5-6, 523-554.

Fama, Eugene F., Lawrence Fisher, Michael C. Jensen, Richard Roll, 1969, The adjustment of stock prices to new information, International Economic Review 10/1, 1-21.

Fatemi, Ali M., 1984, Shareholder Benefits from Corporate International Diversification, Journal of Finance 39/5, 1325-1344.

Federal Deposit Insurance Corporation (FDIC), 08/05/2010, Failed Bank List, http://www.fdic.gov/bank/individual/failed/banklist.html.

Fee, Edward C., Shawn Thomas, 2004, Sources of gains in horizontal mergers: evidence from customer, supplier, and rival firms, Journal of Financial Economics 74/3, 423-460.

Flanagan, Roger, George Norman, 1993, Risk Management and Construction, Blackwell Publishing, n.p.

Flannery, Mark J., 2001, The Faces of "Market Discipline", Journal of Financial Services Research 20/2-3, 107-119.

Ghosh, Chinmoy, Randall Guttery, C. F. Sirmans, 1998, Contagion and REIT Stock Prices, Journal of Real Estate Research 16/3, 389-400.

Ghosh, Chinmoy, Mike Miles, C. F. Sirmans, 1996, Are REITs Stocks?, Real Estate Finance 13, 46-53.

Giannetti, Mariassunta, Andrei Simonov, 2009, On the Real Effects of Bank Bailouts: Micro-Evidence from Japan, Fondazione Eni Enrico Mattei Working Papers, Paper 364.

Giannopoulos, Krzysztof, 1995, Estimating the time-varying components of international stock markets' risk, European Journal of Finance 1/2, 129-164.

Gibbs, Mike, Matthew Webb, Komal Dhillon, 2010, European Beverages, London: J.P.Morgan Cazenove.

Giliberto, Michael S., 1990, Equity Real Estate Investment Trusts and Real Estate Returns, Journal of Real Estate Research 5/2, 259-263.

Goodhart, Charles A. E., Haizhou Huang, 1999, A Model of Lender of Last Resort, IMF working paper No 99-39.

Goold, Michael, Kathleen Luchs, 1993, Why diversify? Four decades of management thinking, Academy of Management Journal 7/3, 7-25.

Gorton, Gary, Lixin Huang, 2004, Liquidity, Efficiency, and Bank Bailouts, The American Economic Review 94/3, 455-484.

Green, Stuart, Scott Fernie, Stephanie Weller, 2005, Making sense of supply chain management: A comparative study of aerospace and construction, Construction Management and Economics 23/6, 579-593.

Greene, William, 2004, The behavior of the maximum likelihood estimator of limited dependent variable models in the presence of fixed effects, Econometrics Journal 7/1, 98-119.

Gropp, Reint, Hendrik Hakenes, Isabel Schnabel, 2011, Competition, Risk-shifting, and Public Bail-out Policies, Review of Financial Studies 24/6, 2084-2120.

Hackbarth, Dirk, Erwan Morellec, 2008, Stock Returns in Mergers and Acquisitions, Journal of Finance 63/3, 1213-1252.

Hadlock, Charles, Michael Ryngaert, Shawn Thomas, 2001, Corporate Structure and Equity Offerings: Are There Benefits to Diversification? Journal of Business 74/4, 613-635.

Hakenes, Hendrik, Isabel Schnabel, 2010, Banks without parachutes: Competitive effects of government bail-out policies, Journal of Financial Stability 6/3, 156-168.

Haldane, Andrew G., Robert M. May, 2011, Systemic risk in banking ecosystems, Nature 469, 351-355.

Hansen, Peter R., Asger Lunde, 2005, A forecast comparison of volatility models: does anything beat a GARCH (1,1)?, Journal of Applied Econometrics 20/7, 873-889.

Harrington, Scott E., 2009, The Financial Crisis, Systematic Risk, and the Future of Insurance Regulation, Journal of Risk and Insurance 76/4, 785-819.

Harrington, Scott E., David G. Shrider, 2007, All Events Induce Variance: Analyzing Abnormal Returns When Effects Vary across Firms, Journal of Financial and Quantitative Analysis 42/1, 229-256.

Haugen, Robert A., Terence C. Langetieg, 1975, An Empirical Test for Synergism in Merger, Journal of Finance 30/4, 1003-1014.

Hawawini, Gabriel A., Itzhak Swary, 1991, Mergers and acquisitions in the U.S. banking industry: Evidence from the capital markets, North-Holland, Amsterdam.

Hentschel, Ludger, 1995, All in the family – Nesting symmetric and asymmetric GARCH models, Journal of Financial Economics 39/1, 71-104.

Horowitz, Anna, Ira Horowitz, 1968, Entropy, Markov Processes and Competition in the Brewing Industry, Journal of Industrial Economics 16/3, 196.

Hull, John C., 2006, Options, Futures, and Other Derivatives, Sixth Edition, Pearson International, 461-478.

Iwasaki, Natsuko, Barry Seldon, Viktor Tremblay, 2008, Brewing Wars of Attrition for Profit (and Concentration), Review of Industrial Organization 33/4, 263-279.

Iyer, Rajkamal, José-Luis Peydró, 2011, Interbank Contagion at Work: Evidence from a Natural Experiment, Review of Financial Studies 24/4, 1337-1377.

Jackson, Leonard A., 2009, Lodging REIT Performance and Comparison With Other Equity REIT Returns, International Journal of Hospitality and Tourism Administration 10, 296-325.

Jagannathan, Ravi, Zhenyu Wang, 1996, The Conditional CAPM and the Cross-Section of Expected Returns, Journal of Finance 51/1, 3-53.

Jensen, Michael C., 2005, Agency costs of overvalued equity costs, Journal of Financial Management 34/1, 5-19.

Jensen, Michael C., William H. Meckling, 1976, Theory of the firm: Managerial behavior, agency costs, and ownership structure, Journal of Financial Economics 3/4, 305-360.

Joehnk, Michael D., James F. Nielsen, 1974, The Effects of Conglomerate Merger Activity on Systematic Risk, Journal of Financial and Quantitative Analysis 9/2, 215-225.

Jones, David, 2010, Top four brewers account for over half world's beer, Reuters UK Web site. Retrieved 7/25/2011, 2011, from http://www.reuters.com/article/2010/02/08/beer-idUSLDE61723K20100208.

Kaufman, George G., 1991, Lender of Last Resort: A Contemporary Perspective, Journal of Financial Services Research 5/2, 95-110.

Kaufman, George G., 1994, Bank Contagion: A Review of the Theory and Evidence, Journal of Financial Services Research 8/2, 123-150.

Kerkvliet, Joe R., William Nebesky, Carol H. Tremblay, Viktor. J. Tremblay, 1998, Efficiency and Technological Change in the U.S. Brewing Industry, Journal of Productivity Analysis 10/3, 271-288.

Kiymaz, Halil, Tarun K. Mukherjee, 2001, Parameter shifts when measuring wealth effects in cross-border mergers, Global Finance Journal 12/2, 249-266.

Klingebiel, Daniela, Randy Kroszner, Luc Laeven, Pieter van Oijen, 2001, Stock Market Responses to Bank Restructuring Policies during the East Asian Crises, World Bank Policy Research Working Paper Series, Paper No. 2571.

Kramer, Lisa A., 2001, Alternative methods for robust analysis in event study applications, Advances in Investment Analysis and Portfolio Management 8, 109-132.

Langetieg, Terence C., Robert A. Haugen, Dean W. Wichern, 1980, Merger and Stockholder Risk, Journal of Financial and Quantitative Analysis 15/03, 689-717.

La Porta, Rafael, Florencio Lopez-De-Silanes, Andrei Shleifer, 2002, Government Ownership of Banks, Journal of Finance 57/1, 265-301.

Lintner, John, 1965, The valuation of risky assets and the selection of risky investments in stock portfolios and capital budgets, Review of Economics and Statistics 47/1, 13-37.

Ljung, Greta M., George E.P. Box, 1978, On a measure of lack of fit in time series models, Biometrica 65/2, 297-303.

Lubatkin, Michael, Sayan Chatterjee, 1994, Extending modern portfolio theory into the domain of corporate diversification: Does it apply?, Academy of Management Journal 37/1, 109-136.

Lubatkin, Michael, Hugh M. O'Neill, 1987, Merger strategies and capital market risk, Academy of Management Journal 30/4, 665-684.

Lynk, Willaim J., 1985, The Price and Output of Beer Revisited, Journal of Business 58/4, 433-437.

Lyon, John D., Brad M. Barber, Chih-Ling Tsai, 1999, Improved Methods for Tests of Long-Run Abnormal Stock Returns, Journal of Finance 54/1, 165-201.

MacKinlay, Craig A., 1997, Event Studies in Economics and Finance, Journal of Economic Literature 35/1, 13-39.

Mandelker, Gershon, 1974, Risk and return: The case of merging firms, Journal of Financial Economics 1/4, 303-335.

Mansi, Sattar A., David M. Reeb, 2002, Corporate Diversification: What Gets Discounted?, Journal of Finance 57/5, 2167-2183.

Martin, John D., Akin Sayrak, 2003, Corporate diversification and shareholder value: A survey of recent literature, Journal of Corporate Finance 9/1, 37-57.

Massey Jr., Frank J., 1951, The Kolmogorov-Smirnov Test for Goodness of Fit, Journal of the American Statistical Association 46/253, 68-78.

Matsusaka, John G., 2001, Corporate diversification, value maximization, and organizational capabilities, Journal of Business 74/3, 409-431.

McAleer, Michael, Felix Chan, Dora Marinova, 2007, An econometric analysis of asymmetric volatility: Theory and application to patents, Journal of Econometrics 139/2, 259-284.

Mehta, Ramit, Dirk Schiereck, 2011, The Consolidation of the Brewing Industry and Wealth Implications from Mergers and Acquisitions, International Journal of Business and Finance Research, forthcoming.

Moeller, Sara B., Frederik P. Schlingemann, 2005, Global diversification and bidder gains: A comparison between cross-border and domestic acquisitions, Journal of Banking and Finance 29/3, 533-564.

Moeller, Sara B., Frederik P. Schlingemann and René M. Stulz, 2003, Do Shareholders of Acquiring Firms Gain from Acquisitions?, NBER working paper No. 9523.

Moeller, Sara B., Frederik P. Schlingemann, René M. Stulz, 2004, Firm size and gains from acquisitions, Journal of Financial Economics 73/2, 201-228.

Montgomery, Cynthia A., Harbir Singh, 1984, Diversification strategy and systematic risk, Strategic Management Journal 5/2, 181-191.

Mueller, Andrew G., Glenn R. Mueller, 2003, Public and Private Real Estate in a Mixed-Asset Portfolio, Journal of Real Estate Portfolio Management 9/3, 193-203.

Nelson, Daniel B., 1991, Conditional heteroscedasticity in asset returns: A new approach, Econometrica 59/2, 347-370.

Newell, Graeme, Franz Fischer, 2009, The Role of Residential REITs in REIT Portfolios, Journal of Real Estate Portfolio Management 15/2, 129-139.

Newell, Graeme, Hsu W. Peng, 2006, The Role of Non-Traditional Real Estate Sectors in REIT Portfolios, Journal of Real Estate Portfolio Management 12/2, 155-166.

Olliges, Jan-Willem, Malte H. Raudszus, Glenn R. Mueller, 2011, Bank Bailouts: REITs and their Performance as Financial Stocks, Working Paper, University of Regensburg.

Payne, James E., 2006, Further evidence on the transmission of shocks across REIT markets: an examination of REIT sub-sectors, Applied Financial Economics Letters 2/3, 141-146.

Payne, James E., Hassan Mohammadi, 2004, The transmission of shocks across real estate investment trust (REIT) markets, Applied Financial Economics 14/7, 1211-1217.

Paulson, Henry M. Jr., 2008, Testimony by Secretary Henry M. Paulson, Jr. before the Senate Banking Committee on Turmoil in US Credit Markets: Recent Actions regarding Government Sponsored Entities, Investment Banks and other Financial Institutions, United States Department of the Treasury, September 23, 2008.

Pettway, Richard H., Takeshi Yamada, 1986, Mergers in Japan and Their Impacts upon Stockholders' Wealth, Financial Management 15/4, 43-52.

Poon, Ser-Huang, Clive W. J. Granger, 2003, Forecasting Volatility in Financial Markets: A Review, Journal of Economic Literature 41/2, 478-539.

Rajan, Raghuram, Henrie Servaes, Luigi Zingales, 2000, The Cost of Diversity: The Diversification Discount and Inefficient Investment, Journal of Finance 55/1, 35-80.

Raudszus, Malte H., Jan-Willem Olliges, Glenn R. Mueller, 2011, Bank Failures & REIT Returns, Working Paper, Tech University of Darmstadt.

Rawley, Evan, 2010, Diversification, Coordination Costs, and Organizational Rigidity: Evidence from Microdata, Strategic Management Journal 31/8, 873-891.

Reinhart, Carmen M., Kenneth S. Rogoff, 2008, Is the 2007 U.S. sub-prime financial crisis so different? An international historical comparison, American Economic Review 98/2, 339-344.

Reinhart, Carmen M., Kenneth S. Rogoff, 2009, The Aftermath of Financial Crises, American Economic Review 99/2, 466-472.

Ro, SeungHan, Alan J. Ziobrowski, 2011, Does Focus Really Matter? Specialized vs. Diversified REITs, Journal of Real Estate Finance and Economics 42/1, 68-83.

Rottke, Nicolas, Dirk Schiereck, Stephan Pauser, 2011, M&A in the Construction Industry -Wealth Effects of Diversification into Real Estate Life Cycle Related Services, International Real Estate Review, 14/3, 283-310.

Scharfstein, David S., 1998, The dark side of internal capital markets II: Evidence from diversified conglomerates, NBER Working Paper W6352, Cambridge, MA.

Schwert, William G., Paul J. Seguin, 1990, Heteroscedasticity in stock returns, Journal of Finance 45/4, 1129-1155.

Shahrur, Husayn, 2005, Industry structure and horizontal takeovers: Analysis of wealth effects on rivals, suppliers, and corporate customers, Journal of Financial Economics 76/1, 61-98.

Sharpe, William F., 1964, Capital asset prices: A theory of market equilibrium under conditions of risk, Journal of Finance 19/3, 425-442.

Slade, Margaret, 2004, Market Power and Joint Dominance in U.K. Brewing, Journal of Industrial Economics 52/1, 133-163.

Song, Moon H., Ralph A. Walkling, 2000, Abnormal returns to rivals of acquisition targets: A test of the 'acquisition probability hypothesis', Journal of Financial Economics 55/2, 143-171.

Stiroh, Kevin J., Adrienne Rumble, 2006, The dark side of diversification: The case of US financial holding companies, Journal of Banking and Finance 30/8, 2131-2161.

Sunder, Shyam, 1980, Stationarity of market risk: Random coefficients test for individual stocks, Journal of Finance 35/4, 883-896.

Tremblay, Viktor J., Carol H. Tremblay, 1988, The Determinants of Horizontal Acquisitions: Evidence from the US Brewing Industry, Journal of Industrial Economics 37/1, 21-45.

U.S. Department of the Treasury (2011): Treasury's bank bailout list: http://www.treasury.gov/initiatives/financial-stability/briefing-room/reports/tarp-transactions/Pages/default.aspx.

Wang, Heli, Jeffrey J. Reuer, 2006, Risk Reduction through Acquisitions: The Role of Firm Specific Investments and Agency Hazards, Advances in Mergers and Acquisitions 5, 25-49.

Wooldridge, Jeffrey M., 1990, A unified approach to robust, regression-based specification tests, Economic Theory 6/1, 17-43.

Zietz, Emily N., Stacy G. Sirmans, Swint H. Friday, 2003, The Environment and Performance of Real Estate Investment Trusts, Journal of Real Estate Portfolio Management 9/2, 127-166.

Appendix

Table A2.1: Sample Overview (I/II): 119 M&A Transactions between 1988 and 2007

Year	N	Horizontal	Vertical	Lateral	Total Transaction Volume (USD mil.)	Average Transaction Volume (USD mil.)
2007	26	11	5	10	14,751.87	567.38
2006	15	3	6	5	3,807.23	271.95
2005	8	4	3	1	8,875.58	1,109.45
2004	5	2	1	2	222.34	44.47
2003	7	6	1	0	2,673.59	381.94
2002	9	5	3	1	981.26	109.03
2001	7	4	2	1	2,234.39	319.20
2000	5	3	2	0	257.68	51.54
1999	6	3	0	3	296.48	49.41
1998	9	5	2	2	323.43	35.94
1997	1	1	0	0	334.80	334.80
1996	7	4	3	0	753.66	107.67
1995	6	6	0	0	331.86	55.31
1994	3	1	2	0	246.25	82.08
1993	1	1	0	0	34.52	34.52
1992	3	2	1	0	105.82	35.27
1989	1	0	0	1	156.20	156.20
1988	1	0	1	0	31.94	31.94
Sum	119	61	32	26	36,418.89	188.90

The sample includes all M&A transactions between 1988 and 2007, with acquirers classified as within the construction industry (SIC codes >= 1,500 and < 1,800), and all other listed constraints were fulfilled. The table shows cumulative M&As per year, with total and average transaction values in millions of USD. Additionally, the number of transactions in three subgroups is given by transaction direction. If the target is also classified as within the construction industry, the transaction direction is horizontal, if the target is not within the construction industry but is in an industry that is related to it according to value chain principles, it is considered vertical. All other cases are classified as lateral M&A transactions.

Table A2.2: Sample Overview (II/II): M&As along Acquirers' and Targets' Countries of Origin

	Acquir.	Europe												NA			Asia							RoW			Sum
	Targets	CK	GB	GE	GR	IT	FR	PD	PG	NE	SP	SW	TY	US	CA	PA	JN	SK	SI	MA	RU	TW	CH	AU	NZ	IL	
Europe	CK	1									1																2
	GB		28				1				1																30
	GE			1						1																	2
	GR				1																						1
	IT					4																					4
	FR						6																				6
	PD						1	1																			2
	PG								1		1																2
	NE									1																	1
	SP			1							6			1													8
	SW											2															2
NA	US		2	1							2			26	2	1									1		35
	CA		1											1	1												3
Asia	JN																3										3
	SK																	1									1
	MA																			3							3
	RU												1								1						2
	TW																					1					1
	CH																						1				1
	HK			1															1								2
RoW	AU			1																				3	1		5
	IL																									1	1
	BR													1													1
	AR										1																1
Sum		1	31	5	1	4	8	1	1	2	12	2	1	29	3	1	3	1	1	3	1	1	1	3	2	1	119

The sample has been clustered to the world's three major economic regions: Europe, North America (NA), and Asia. The remaining transactions are classified as the "rest of world" (RoW) category. Country abbreviations are as follows: CK: Czech Republic, GB: Great Britain, GE: Germany, GR: Greece, IT: Italy, FR: France, PD: Poland, PG: Portugal, NE: Netherlands, SP: Spain, SW: Sweden, TY: Turkey, US: United States of America, CA: Canada, PA: Panama, JN: Japan, SK: South Korea, SI: Singapore, MA: Malaysia, RU: Russia, TW: Taiwan, CH: China, HK: Hong Kong, AU: Australia, NZ: New Zealand, IL: Israel, BR: Brazil, and AR: Argentina.

Table A2.3: Standard Event Study: Construction M&As along Transaction Volume

CAAR	[-x; +y]		Mean	Median		KS-test	p	Wx-test	p	Z-stat.	p	t-test	p	n
Low <40mU$D	0	0	1.94%	1.53%	***	0.12	0.54	-3.01	0.00	3.57	0.00	3.58	0.01	39
	0	1	2.79%	1.31%	***	0.13	0.44	-3.15	0.00	3.56	0.00	3.70	0.01	39
	0	2	2.78%	1.72%	***	0.12	0.55	-2.44	0.01	2.83	0.00	3.07	0.01	39
	0	7	1.58%	1.17%		0.11	0.67	-1.34	0.18	1.23	0.22	1.52	0.05	39
	0	10	1.37%	1.41%		0.15	0.28	-0.96	0.34	1.06	0.29	1.27	0.10	39
Medium <130mU$D	0	0	0.21%	-0.01%		0.21	0.05	-0.28	0.78	0.04	0.97	0.20	0.50	40
	0	1	0.98%	0.10%		0.18	0.12	-0.86	0.39	0.80	0.43	0.83	0.50	40
	0	2	0.82%	-0.36%		0.15	0.30	-0.17	0.86	0.59	0.56	0.74	0.50	40
	0	7	0.16%	0.81%		0.18	0.12	-0.85	0.40	0.21	0.84	0.10	0.50	40
	0	10	-0.25%	0.73%		0.16	0.24	-0.15	0.88	-0.05	0.96	-0.15	0.50	40
High >130mU$D	0	0	0.76%	1.18%		0.14	0.37	-1.13	0.26	1.07	0.28	1.16	0.10	40
	0	1	1.20%	0.81%	**	0.13	0.48	-2.00	0.05	1.72	0.09	1.82	0.05	40
	0	2	1.34%	0.50%		0.11	0.68	-1.20	0.23	1.66	0.10	1.53	0.05	40
	0	7	1.39%	0.39%		0.11	0.63	-1.13	0.26	1.69	0.09	1.36	0.05	40
	0	10	-0.15%	-0.39%		0.09	0.88	-0.01	0.99	0.24	0.81	-0.17	0.50	40

***, **, and * denote statistical significance at the 1%, 5%, and 10% levels, respectively. *CAAR*: cumulative average abnormal return, KS-test: Kolmogorov-Smirnov test - p-value ≤ 0.05 indicates normal distribution at the 5% significance level, Wx-test: Wilcoxon signed rank test as a non-parametric median test, Z-stat.: Boehmer test (parametric), t-test: skewness adjusted and bootstrapped t-test.

Table A2.4: Risk Event Study: Construction M&As along Transaction Volume

CAAB	[-x; +y]		Mean	Median		KS-test	P	Wx-test	p	t-test	p	n
Low <40mU$D	0	0	0.04	0.02		0.36	0.00	-1.51	0.13	1.27	0.21	39
	0	1	0.09	0.04		0.39	0.00	-1.60	0.11	1.34	0.19	39
	0	2	0.18	0.07	+	0.40	0.00	-1.65	0.10	1.33	0.19	39
	0	7	0.67	0.12	+	0.38	0.00	-1.66	0.10	1.26	0.22	39
	0	10	0.96	0.18		0.38	0.00	-1.54	0.12	1.24	0.22	39
Medium <130mU$D	0	0	-0.02	0.00		0.19	0.10	-1.18	0.24	-1.28	0.21	40
	0	1	-0.04	-0.01		0.20	0.06	-0.97	0.33	-1.02	0.31	40
	0	2	-0.05	-0.01		0.21	0.05	-0.94	0.35	-0.91	0.37	40
	0	7	-0.11	-0.03		0.21	0.05	-0.67	0.50	-0.63	0.53	40
	0	10	-0.12	-0.03		0.18	0.12	-0.46	0.65	-0.48	0.63	40
High ≥130mU$D	0	0	0.01	0.00		0.16	0.24	-0.16	0.87	0.73	0.47	40
	0	1	0.02	0.01		0.17	0.16	-0.01	0.99	0.54	0.59	40
	0	2	0.02	0.01		0.19	0.09	-0.20	0.84	0.39	0.70	40
	0	7	0.02	0.02		0.15	0.28	-0.28	0.78	0.17	0.87	40
	0	10	0.03	-0.05		0.16	0.21	-0.38	0.71	0.15	0.88	40

+ indicates statistical significance at the 10% levels according to Wilcoxon signed rank test, even though normal distribution may be assumed and t-test should return appropriate significance values. *CAAB*: cumulative average abnormal beta shift, KS-test: Kolmogorov-Smirnov test - p-value ≤ 0.05 indicates normal distribution at the 5% significance level, Wx-test: Wilcoxon signed rank test as a non-parametric median test, t-test: standard parametric mean test.

Table A2.5: Long-term Risk Analysis: Equity Beta Dynamics along Transaction Volume

Equity beta change in years relative to event												
Low	[-x; +y]		Mean	Median		KS-test	p	Wx-test	p	t-test	p	n
	-1	1	0.02	0.04		0.11	0.74	-0.89	0.37	0.42	0.68	39
OLS	-1	2	0.09	0.15	*	0.10	0.75	-1.66	0.10	1.21	0.23	39
	1	2	0.07	0.08	*	0.12	0.62	-1.72	0.09	1.28	0.21	39
	-1	1	0.01	0.11		0.11	0.66	-0.75	0.45	0.19	0.85	39
GARCH	-1	2	0.17	0.11		0.11	0.74	-1.58	0.11	1.56	0.13	39
	1	2	0.16	0.11		0.18	0.15	-1.62	0.11	1.82	0.08	39
Medium												
	-1	1	0.02	0.01		0.14	0.40	-0.52	0.60	0.47	0.64	40
OLS	-1	2	-0.09	-0.08		0.12	0.59	-1.18	0.24	-1.15	0.26	40
	1	2	-0.11	-0.06	*	0.10	0.77	-1.65	0.10	-1.90	0.06	40
	-1	1	-0.06	0.02		0.14	0.36	-0.05	0.96	-0.73	0.47	40
GARCH	-1	2	-0.10	-0.13		0.15	0.28	-0.75	0.45	-0.64	0.53	40
	1	2	-0.04	-0.05		0.14	0.35	-0.48	0.63	-0.38	0.71	40
High												
	-1	1	0.03	0.06		0.10	0.82	-0.33	0.74	0.43	0.67	39
OLS	-1	2	0.01	-0.05		0.09	0.90	-0.06	0.96	0.09	0.93	39
	1	2	-0.02	0.02		0.10	0.76	-0.35	0.73	-0.48	0.63	39
	-1	1	0.02	0.05		0.16	0.25	-0.46	0.65	0.14	0.89	39
GARCH	-1	2	0.01	0.19		0.15	0.32	-0.93	0.35	0.08	0.94	39
	1	2	-0.01	0.03		0.13	0.47	-0.42	0.68	-0.08	0.94	39

***, **, and * denote statistical significance at the 1%, 5%, and 10% levels, respectively. Subsample definition: Low: transaction volume (TV) < USD 40 million, medium: TV ≥ USD 40 million but < USD 130 million, high: TV ≥ USD 130 million, KS-test: Kolmogorov-Smirnov test - p-value ≤ 0.05 indicates normal distribution at the 5% significance level, Wx-test: Wilcoxon signed rank test as a non-parametric median test, t-test: standard parametric mean test.

Table A2.6: Long-term Risk Analysis: Equity/Asset Beta across Economic Cycles

Equity beta change in years relative to event											
1996-'01	[-x; +y]	Mean	Median		KS-test	p	Wx-test	p	t-test	p	n
	-1 1	-0.01	0.00		0.12	0.69	-0.26	0.79	-0.12	0.91	35
OLS	-1 2	-0.05	-0.05		0.19	0.12	-1.36	0.17	-0.70	0.49	35
	1 2	-0.05	0.00		0.16	0.29	-0.46	0.65	-0.76	0.45	35
	-1 1	-0.08	-0.06		0.06	1.00	-1.42	0.15	-1.38	0.18	35
GARCH	-1 2	0.02	-0.03		0.17	0.26	-0.51	0.61	0.24	0.81	35
	1 2	0.10	0.02		0.10	0.80	-0.90	0.37	1.17	0.25	35
2002-'07											
	-1 1	0.07	0.07		0.08	0.79	-1.50	0.13	1.53	0.13	68
OLS	-1 2	0.09	0.12	**	0.10	0.51	-1.92	0.05	1.65	0.10	68
	1 2	0.02	0.07		0.09	0.56	-1.00	0.32	0.40	0.69	68
	-1 1	0.06	0.12	+	0.16	0.05	-2.29	0.02	0.81	0.42	68
GARCH	-1 2	0.09	0.23	+	0.16	0.04	-2.13	0.03	0.85	0.40	68
	1 2	0.03	0.07		0.13	0.20	-0.74	0.46	0.42	0.67	68
Asset beta change in years relative to event											
1996-'01											
	-1 1	-0.06	-0.08		0.13	0.64	-1.61	0.11	-1.26	0.22	29
OLS	-1 2	-0.07	-0.05	*	0.16	0.42	-1.85	0.06	-1.16	0.25	29
	1 2	-0.01	0.00		0.13	0.65	-0.18	0.85	-0.21	0.84	29
	-1 1	-0.12	-0.14	**	0.07	0.99	-2.26	0.02	-2.29	0.03	29
GARCH	-1 2	-0.04	-0.02		0.09	0.96	-0.88	0.38	-0.67	0.51	29
	1 2	0.08	0.03		0.08	0.98	-1.39	0.16	1.46	0.15	29
2002-'07											
	-1 1	0.03	0.00		0.12	0.32	-0.40	0.69	0.65	0.52	60
OLS	-1 2	0.02	0.05		0.09	0.67	-0.85	0.40	0.44	0.66	60
	1 2	-0.01	0.03		0.11	0.47	-0.07	0.95	-0.24	0.81	60
	-1 1	0.01	0.06		0.18	0.04	-1.46	0.14	0.14	0.89	60
GARCH	-1 2	0.00	0.11	+	0.24	0.00	-1.94	0.05	-0.02	0.98	60
	1 2	-0.01	0.02		0.13	0.26	-0.24	0.81	-0.25	0.80	60

***, **, and * denote statistical significance at the 1%, 5%, and 10% levels, respectively. + denotes statistical significance at a minimum of the 10% level, according to the Wilcoxon signed rank test, even though normal distributions may be assumed, and a t-test should return appropriate significance values. We do not include the 1988-1995 period here, because of its minimal amount of transactions (fifteen) for this economic cycle. KS-test: Kolmogorov-Smirnov test for normal distributions, Wx-test: Wilcoxon signed rank test as a non-parametric median test, t-test: standard parametric mean test.

Table A2.7: Multivariate Short-term Risk Analysis

Dependent variable CAB [-10;+10]										
Model		1	2	3	4	5	6	7	8	9
	α	0.08 *	0.00	0.01	0.00	0.00	0.00	0.02	0.00	0.02
		1.7	*0.2*	*1.0*	*0.3*	*0.4*	*0.1*	*0.9*	*-0.2*	*1.3*
Independent variables										
Lateral M&A	δ_1	0.01	0.01							
		0.6	*0.7*							
Vertical M&A	δ_2	-0.02		-0.03						
		-0.9		*-1.2*						
Transaction volume (TV)	δ_3	0.00 *			0.00					
		1.6			*1.3*					
Relative size: TV/acq. MV	δ_4	0.00				0.00				
		-0.1				*0.6*				
1996-2001	δ_5	-0.03					0.01			
		-0.6					*0.4*			
2002-2007	δ_6	-0.05						-0.03		
		-1.5						*-1.1*		
Europe	δ_7	-0.04							0.01	
		-1.2							*0.6*	
North America	δ_8	-0.07 **								-0.04 *
		-2.0								*-1.7*
Nobs.		119								
R^2 (adj.)		0.00	-0.01	0.00	-0.01	-0.01	-0.01	0.00	-0.01	0.02
F-stat. (p-value)		0.57	0.89	0.48	0.91	1.00	0.87	0.44	0.84	0.24
DW statistic		1.76	1.78	1.76	1.80	1.80	1.80	1.82	1.79	1.77

CAB: cumulative abnormal beta shift. ***, **, and * denote statistical significance at the 1%, 5%, and 10% levels, respectively; t-statistics in italics.

Table A2.8: Multivariate Long-term Risk Analysis: GARCH Equity Beta Changes

Dependent variable GARCH equity beta change [-1;+2]										
Model		1	2	3	4	5	6	7	8	9
	α	-0.22	-0.01	0.02	0.03	0.03	0.03	-0.06	-0.04	0.09
		-1.0	*-0.1*	*0.3*	*0.4*	*0.4*	*0.3*	*-0.8*	*-0.3*	*1.4*
Independent variables										
Lateral M&A	δ_1	0.18	0.17							
		1.1	*1.2*							
Vertical M&A	δ_2	0.08		0.02						
		0.4		*0.1*						
Transaction volume (TV)	δ_3	0.00			0.00					
		-0.4			*-0.2*					
Relative size: (TV/acq. MV)	δ_4	-0.01 ***				0.00 **				
		-2.5				*-2.2*				
1996-2001	δ_5	0.28 *					0.00			
		1.8					*0.0*			
2002-2007	δ_6	0.35 **						0.16		
		2.1						*1.1*		
Europe	δ_7	-0.01							0.11	
		-0.1							*0.7*	
North America	δ_8	-0.27								-0.24
		-1.2								*-1.1*
Nobs.		118								
R^2 (adj.)		-0.02	0.00	-0.01	-0.01	-0.01	0.00	0.00	0.00	0.01
F-stat. (p-value)		0.70	0.60	0.99	0.97	0.97	1.00	0.55	0.77	0.36
DWstatistic		1.65	1.60	1.62	1.63	1.63	1.63	1.64	1.61	1.62

***, **, and * denote statistical significance at the 1%, 5%, and 10% levels, respectively; t-statistics in italics.

Table A3.1: Acquirer CAABs of Overall Transaction Sample

CAAB [-x; +y]	Mean		Median	KS-test	p	Wx-test	p	t-test	p	N
0 0	-0.01		0.00	0.12	0.23	-1.03	0.30	-0.02	0.32	75
1 1	-0.02		0.00	0.10	0.36	-1.24	0.22	-0.06	0.22	75
2 2	-0.04		-0.01	0.10	0.39	-1.29	0.20	-0.10	0.18	75
5 5	-0.09		-0.03	0.10	0.44	-1.39	0.16	-0.23	0.18	75
10 10	-0.15		-0.02	0.09	0.51	-1.30	0.19	-0.40	0.23	75
20 20	-0.24		-0.13	0.10	0.38	-1.29	0.20	-0.68	0.29	75
0 1	-0.02		0.00	0.13	0.13	-1.25	0.21	-0.04	0.16	75
0 2	-0.03		-0.01	0.13	0.17	-1.48	0.14	-0.07	0.11	75
0 5	-0.07	*	-0.03	0.12	0.21	-1.61	0.11	-0.16	0.10	75
0 10	-0.12		-0.08	0.10	0.47	-1.44	0.15	-0.29	0.15	75
0 20	-0.18		-0.05	0.08	0.64	-1.19	0.23	-0.52	0.28	75

***, **, and * indicate statistical significance at the 1%, 5%, and 10% levels, respectively. *CAAB*: cumulative average abnormal beta shift, KS test: Kolmogorov-Smirnov test - p-value ≤ 0.05 indicates normal distribution at the 5% significance level, Wx-test: Wilcoxon signed rank test as non-parametric median test, t-test: std. parametric mean test.

Table A3.2: Acquirer CAABs of Domestic Transactions

CAAB [-x; +y]	Mean	Median	KS-test	p	Wx-test	p	t-test	p	N
0 0	0.00	0.00	0.23	0.15	-0.46	0.65	-0.03	0.81	24
1 1	-0.01	0.01	0.23	0.14	-0.69	0.49	-0.08	0.86	24
2 2	-0.01	0.01	0.24	0.11	-0.49	0.63	-0.13	0.86	24
5 5	-0.02	0.02	0.24	0.11	-0.57	0.57	-0.28	0.86	24
10 10	0.02	0.10	0.19	0.29	-0.89	0.38	-0.46	0.94	24
20 20	0.16	0.11	0.15	0.57	-0.89	0.38	-0.68	0.71	24
0 1	-0.01	0.00	0.24	0.11	-0.69	0.49	-0.06	0.77	24
0 2	-0.01	0.01	0.23	0.13	-0.34	0.73	-0.09	0.75	24
0 5	-0.02	0.01	0.21	0.22	-0.20	0.84	-0.18	0.78	24
0 10	0.01	0.02	0.16	0.55	-0.57	0.57	-0.28	0.92	24
0 20	0.16	0.04	0.16	0.55	-0.80	0.42	-0.43	0.59	24

***, **, and * indicate statistical significance at the 1%, 5%, and 10% levels, respectively. *CAAB*: cumulative average abnormal beta shift, KS test: Kolmogorov-Smirnov test - p-value ≤ 0.05 indicates normal distribution at the 5% significance level, Wx-test: Wilcoxon signed rank test as non-parametric median test, t-test: std. parametric mean test.

Table A3.3: Acquirer CAABs by Relative Size of Transaction

CAAB	[-x; +y]	Mean	Median		KS-test	p	Wx-test	p	t-test	p	n
Bottom 30 (<6%)	0 0	-0.01	-0.01	*	0.17	0.33	-1.68	0.09	-0.03	0.26	30
	1 1	-0.03	-0.03		0.15	0.44	-1.53	0.13	-0.09	0.29	30
	2 2	-0.05	-0.05		0.15	0.48	-1.51	0.13	-0.15	0.28	30
	5 5	-0.12	-0.11		0.14	0.55	-1.57	0.12	-0.33	0.22	30
	10 10	-0.23	-0.22		0.15	0.45	-1.57	0.12	-0.61	0.21	30
	20 20	-0.44	-0.51	*	0.12	0.74	-1.64	0.10	-1.06	0.16	30
	0 1	-0.02	-0.02		0.16	0.41	-1.55	0.12	-0.06	0.25	30
	0 2	-0.04	-0.03	*	0.14	0.57	-1.66	0.10	-0.10	0.22	30
	0 5	-0.08	-0.07	*	0.15	0.50	-1.82	0.07	-0.20	0.16	30
	0 10	-0.16	-0.14	*	0.12	0.69	-1.66	0.10	-0.38	0.13	30
	0 20	-0.32	-0.28	*	0.10	0.92	-1.68	0.09	-0.74	0.13	30
Top 30 (19%><430%)	0 0	0.00	0.00		0.11	0.85	-0.20	0.85	-0.01	0.94	30
	1 1	-0.02	0.00		0.12	0.74	-0.42	0.67	-0.07	0.51	30
	2 2	-0.04	-0.01		0.13	0.65	-0.59	0.56	-0.12	0.39	30
	5 5	-0.09	-0.04		0.13	0.69	-0.73	0.47	-0.30	0.35	30
	10 10	-0.18	-0.10		0.13	0.63	-0.75	0.45	-0.56	0.36	30
	20 20	-0.26	-0.17		0.12	0.70	-0.65	0.52	-0.96	0.46	30
	0 1	-0.02	0.00		0.15	0.50	-0.36	0.72	-0.05	0.41	30
	0 2	-0.03	-0.01		0.16	0.37	-0.71	0.48	-0.10	0.30	30
	0 5	-0.09	-0.02		0.16	0.42	-0.89	0.37	-0.24	0.25	30
	0 10	-0.15	-0.07		0.15	0.50	-0.89	0.37	-0.45	0.30	30
	0 20	-0.21	-0.04		0.15	0.48	-0.59	0.56	-0.80	0.47	30

***, **, and * indicate statistical significance at the 1%, 5%, and 10% levels, respectively. *CAAB*: cumulative average abnormal beta shift, KS test: Kolmogorov-Smirnov test - p-value ≤ 0.05 indicates normal distribution at the 5% significance level, Wx-test: Wilcoxon signed rank test as non-parametric median test, t-test: std. parametric mean test.

Table A3.4: Acquirer CAABs by Date of Transaction Announcement

CAAB	[-x; +y]	Mean	Median	KS-test	p	Wx-test	p	t-test	p	n
1991 - 2003	0 0	-0.01	0.00	0.12	0.64	-0.88	0.38	-0.02	0.31	36
	1 1	-0.02	-0.01	0.10	0.82	-0.86	0.39	-0.07	0.34	36
	2 2	-0.04	-0.01	0.09	0.89	-0.90	0.37	-0.12	0.32	36
	5 5	-0.08	-0.03	0.11	0.78	-1.04	0.30	-0.25	0.35	36
	10 10	-0.12	-0.01	0.12	0.64	-0.88	0.38	-0.44	0.45	36
	20 20	-0.18	0.00	0.17	0.22	-0.79	0.43	-0.76	0.54	36
	0 1	-0.02	0.00	0.11	0.74	-0.93	0.35	-0.05	0.27	36
	0 2	-0.03	-0.01	0.10	0.80	-1.02	0.31	-0.08	0.25	36
	0 5	-0.06	-0.02	0.12	0.67	-1.27	0.20	-0.16	0.26	36
	0 10	-0.08	-0.04	0.12	0.66	-1.05	0.29	-0.29	0.42	36
	0 20	-0.12	-0.03	0.15	0.34	-0.64	0.52	-0.54	0.58	36
2004 - 2010	0 0	0.00	0.00	0.16	0.27	-0.64	0.52	-0.02	0.65	39
	1 1	-0.02	0.00	0.13	0.46	-0.92	0.36	-0.08	0.42	39
	2 2	-0.04	0.00	0.12	0.54	-0.89	0.37	-0.14	0.36	39
	5 5	-0.10	-0.02	0.12	0.54	-0.93	0.35	-0.32	0.33	39
	10 10	-0.18	-0.15	0.11	0.72	-0.91	0.36	-0.57	0.36	39
	20 20	-0.29	-0.47	0.12	0.64	-0.92	0.36	-0.99	0.40	39
	0 1	-0.02	0.00	0.15	0.28	-0.84	0.40	-0.06	0.35	39
	0 2	-0.04	-0.01	0.15	0.29	-1.00	0.32	-0.10	0.27	39
	0 5	-0.09	-0.06	0.15	0.33	-1.12	0.26	-0.23	0.22	39
	0 10	-0.16	-0.12	0.13	0.49	-0.98	0.33	-0.43	0.24	39
	0 20	-0.25	-0.27	0.09	0.86	-0.84	0.40	-0.78	0.36	39

***, **, and * indicate statistical significance at the 1%, 5%, and 10% levels, respectively. *CAAB*: cumulative average abnormal beta shift, KS test: Kolmogorov-Smirnov test - p-value ≤ 0.05 indicates normal distribution at the 5% significance level, Wx-test: Wilcoxon signed rank test as non-parametric median test, t-test: std. parametric mean test.

Table A3.5: Rival CAABs by Transaction Value

CAAB	[-x; +y]	Mean	Median		KS-test	p	Wx-test	p	t-test	p	n
Bottom 30 (50M><132M)	0 0	0.01	0.01	**	0.14	0.55	-2.01	0.04	0.00	0.21	30
	1 1	0.02	0.02	*	0.13	0.69	-1.64	0.10	-0.02	0.34	30
	2 2	0.02	0.03		0.16	0.38	-1.61	0.11	-0.04	0.51	30
	5 5	0.03	0.05		0.19	0.18	-1.47	0.14	-0.12	0.67	30
	10 10	0.06	0.09		0.17	0.31	-1.20	0.23	-0.22	0.66	30
	20 20	0.11	0.15		0.18	0.26	-1.18	0.24	-0.39	0.66	30
	0 1	0.01	0.02	*	0.14	0.56	-1.70	0.09	-0.01	0.32	30
	0 2	0.01	0.02		0.17	0.29	-1.51	0.13	-0.03	0.54	30
	0 5	0.02	0.04		0.16	0.37	-1.31	0.19	-0.07	0.65	30
	0 10	0.05	0.05		0.14	0.52	-1.31	0.19	-0.13	0.56	30
	0 20	0.08	0.06		0.19	0.22	-1.20	0.23	-0.32	0.69	30
Top 30 (646M><53B)	0 0	0.01	0.01		0.11	0.79	-1.35	0.18	-0.01	0.20	30
	1 1	0.04	0.03		0.10	0.89	-1.39	0.17	-0.02	0.19	30
	2 2	0.07	0.04		0.11	0.84	-1.45	0.15	-0.03	0.18	30
	5 5	0.14	0.07		0.11	0.83	-1.39	0.17	-0.07	0.19	30
	10 10	0.22	0.16		0.13	0.66	-1.33	0.18	-0.16	0.25	30
	20 20	0.33	0.39		0.14	0.58	-1.12	0.26	-0.34	0.32	30
	0 1	0.03	0.02		0.11	0.84	-1.49	0.14	-0.01	0.16	30
	0 2	0.04	0.03		0.12	0.76	-1.47	0.14	-0.02	0.17	30
	0 5	0.08	0.07		0.10	0.90	-1.24	0.21	-0.05	0.21	30
	0 10	0.10	0.18		0.11	0.85	-1.02	0.31	-0.14	0.41	30
	0 20	0.11	0.32		0.14	0.57	-0.87	0.38	-0.36	0.63	30

***, **, and * indicate statistical significance at the 1%, 5%, and 10% levels, respectively. *CAAB*: cumulative average abnormal beta shift, KS test: Kolmogorov-Smirnov test - p-value ≤ 0.05 indicates normal distribution at the 5% significance level, Wx-test: Wilcoxon signed rank test as non-parametric median test, t-test: std. parametric mean test.

Table A3.6: Rival CAABs by Date of Transaction Announcement

CAAB	[-x; +y]	Mean	Median		KS-test	p	Wx-test	p	t-test	p	n
1991 - 2003	0 0	0.01	0.01	**	0.15	0.34	-2.04	0.04	0.00	0.06	37
	1 1	0.04	0.02	**	0.15	0.36	-2.01	0.04	0.00	0.06	37
	2 2	0.06	0.03	*	0.12	0.66	-1.88	0.06	0.00	0.05	37
	5 5	0.14	0.06	*	0.12	0.65	-1.70	0.09	-0.01	0.06	37
	10 10	0.25	0.09	*	0.13	0.53	-1.70	0.09	-0.01	0.05	37
	20 20	0.52	0.35	**	0.11	0.73	-2.07	0.04	0.06	0.03	37
	0 1	0.03	0.01	**	0.14	0.41	-2.23	0.03	0.00	0.05	37
	0 2	0.04	0.02	**	0.13	0.53	-1.98	0.05	0.00	0.06	37
	0 5	0.07	0.04	*	0.09	0.86	-1.68	0.09	-0.01	0.10	37
	0 10	0.12	0.05		0.08	0.94	-1.50	0.13	-0.03	0.12	37
	0 20	0.28	0.18		0.08	0.94	-1.59	0.11	-0.04	0.08	37
2004 - 2010	0 0	0.00	0.01		0.20	0.02	-1.09	0.28	-0.02	0.79	59
	1 1	-0.01	0.02		0.19	0.02	-0.95	0.34	-0.05	0.73	59
	2 2	-0.02	0.03		0.18	0.04	-0.91	0.37	-0.09	0.66	59
	5 5	-0.04	0.05		0.20	0.02	-0.82	0.41	-0.20	0.58	59
	10 10	-0.10	0.09		0.20	0.02	-0.60	0.55	-0.40	0.49	59
	20 20	-0.22	0.12		0.20	0.02	-0.32	0.75	-0.75	0.42	59
	0 1	0.00	0.01		0.19	0.03	-0.93	0.35	-0.03	0.77	59
	0 2	-0.01	0.02		0.17	0.06	-0.84	0.40	-0.05	0.71	59
	0 5	-0.02	0.04		0.16	0.07	-0.66	0.51	-0.11	0.66	59
	0 10	-0.05	0.04		0.14	0.15	-0.40	0.69	-0.22	0.56	59
	0 20	-0.12	0.03		0.17	0.05	-0.07	0.95	-0.45	0.45	59

***, **, and * indicate statistical significance at the 1%, 5%, and 10% levels, respectively. *CAAB*: cumulative average abnormal beta shift, KS test: Kolmogorov-Smirnov test - p-value ≤ 0.05 indicates normal distribution at the 5% significance level, Wx-test: Wilcoxon signed rank test as non-parametric median test, t-test: std. parametric mean test.

Table A4.1: Descriptive Statistics: Bank Failures Sample

Sample characteristics	*Total*	*Min*	*Max*	*Mean*	*STD*
Date vector	11.5 years	Jan-99	Jul-10	-	-
Bank failures	308	0/year*	140/year**	25.7/year	46
Bank failure dates	130	0/year*	43/year**	10.8/year	13
Failures per event date	-	1	9	2.4	2.0
Days between failure dates	-	1	952	32	89
Failure dates on Fridays	112	-	-	86.2%	-
Assets at failure dates (USD)	2.2T	10.2M	1.6T	17.2B	143B
Deposits at failure dates (USD)	446B	0***	188B	3.4B	18B
FDIC costs at failure dates (USD)	73B	0***	8.9B	561M	1.2B

*)2005, 2006; **) 2009, extrapolating for full year 2010: no. of bank failure dates »46, no. of bank failures »175; ***) Lehman Brothers had no deposits due to its investment bank status.

Figure A4.1: Failed Banks' Total Assets Cumulated per Year (01/1999 – 07/2010)

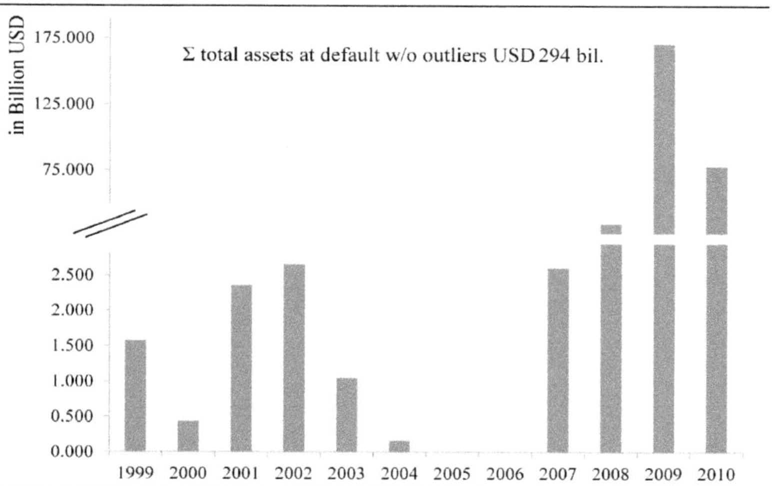

Comment: Number of bank failures: 305; number of bank failure dates: 127 (outliers excluded)
Source: illustration by authors.

Figure A4.2: Failed Banks' Total Assets Distribution per Year (01/1999 – 12/2004)

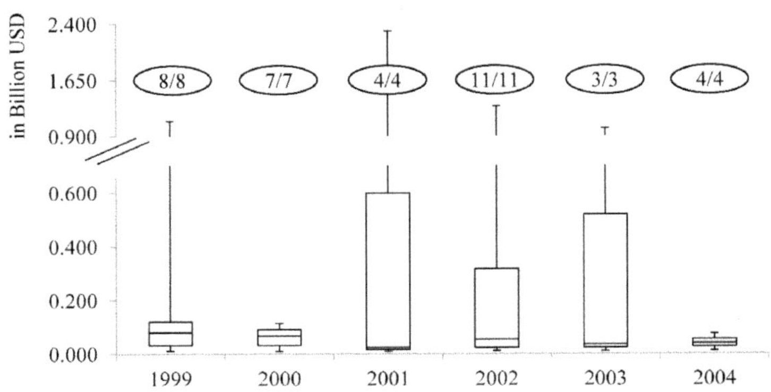

Illustration: First figure in bubble indicates number of bank failures, second figure indicates number of bank failure dates
Comment: Number of bank failures: 37; number of bank failure dates: 37
Source: illustration by authors.

Figure A4.3: Failed Banks' Total Assets Distribution per Year (01/2007 – 07/2010)

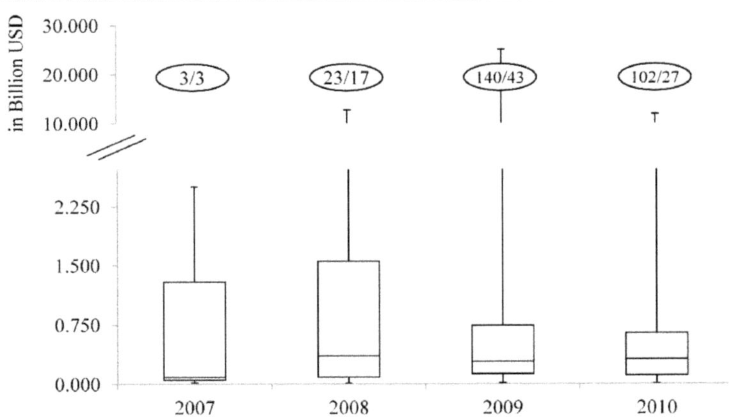

Illustration: First figure in bubble indicates number of bank failures, second figure indicates number of bank failure dates
Comment: Number of bank failures: 268; number of bank failure dates: 90 (outliers excluded)
Source: illustration by authors.

Table A4.2: Descriptive Statistics: REIT and Market Return Data

Return Indices	*Short description*	*available from**
S&P 500 Composite	U.S. market index (common equities)	1/1/1998
FTSE EPRA/NAREIT US	U.S. all REIT index	1/1/1998
DJ Equity all REIT	U.S. equity REIT index	9/1/2000
US-DSMortgage REITs	U.S. mortgage REIT index	10/8/1997
MSCI US Industrial REITs	U.S. industrial REIT index	8/7/2006
US-DSHotel, Ldg REITs	U.S. lodging REIT index	1/1/1990
S&P1500 Office REITs	U.S. office REIT index	4/28/2006
S&P1500 Retail REITs	U.S. retail REIT index	4/28/2006
S&P1500 Residential REITs	U.S. residential REIT index	4/28/2006
FTSE EPRA/NAREIT US Healthcare	U.S. healthcare REIT index	1/18/2007
AMERICAS-DS Specialty REITs	U.S. specialty REIT index	1/1/1990
FTSE EPRA/NAREIT Dev NA	U.S. diversified REIT index	1/1/1990

*) concerns the availability of return data with respect to our selected time period - earliest date is thus given as 1/1/1998.

Table A6.1: EGARCH (1,1) Estimation of Subindustry Indices for 2007-2010

Financial Services

	REIT	BANKS	INSUR	ASSET MGR	BROKERS
γ_1	-0.09 ***	-0.13 ***	-0.15 ***	-0.10 ***	-0.20 ***
	-2.8	*-3.8*	*-4.3*	*-3.2*	*-4.4*
γ_2	0.99 ***	0.98 ***	0.98 ***	0.99 ***	0.97 ***
	212.3	*218.9*	*234.5*	*239.2*	*174.1*
γ_3	0.18 ***	0.29 ***	0.22 ***	0.18 ***	0.20 ***
	5.8	*8.6*	*7.0*	*7.5*	*7.1*
γ_4	-0.08 ***	-0.08 ***	-0.10 ***	-0.12 ***	-0.09 ***
	-3.5	*-4.7*	*-4.5*	*-5.7*	*-4.5*

Industrials

	CON&MAT	AER&DEF	GEN IND	ELE EQU	IND ENG
γ_1	-0.13 ***	-0.18 ***	-0.14 ***	-0.13 ***	-0.13 ***
	-2.7	*-4.1*	*-3.4*	*-3.5*	*-3.4*
γ_2	0.98 ***	0.98 ***	0.98 ***	0.98 ***	0.98 ***
	157.2	*183.6*	*184.4*	*221.6*	*199.4*
γ_3	0.17 ***	0.14 ***	0.15 ***	0.14 ***	0.15 ***
	5.4	*4.7*	*5.9*	*6.2*	*5.0*
γ_4	-0.07 ***	-0.11 ***	-0.08 ***	-0.09 ***	-0.08 ***
	-3.7	*-5.3*	*-4.5*	*-5.1*	*-4.3*

Consumer Goods

	AUT&PAR	BEVER	FD PROD	HOU GDS	LEI GDS
γ_1	-0.12 **	-0.30 ***	-0.14 ***	-0.19 ***	-0.22 ***
	-2.3	*-3.8*	*-3.9*	*-3.7*	*-4.3*
γ_2	0.98 ***	0.97 ***	0.98 ***	0.98 ***	0.97 ***
	150.5	*111.0*	*244.5*	*162.9*	*148.0*
γ_3	0.17 ***	0.19 ***	0.11 ***	0.16 ***	0.14 ***
	5.2	*5.8*	*4.2*	*6.0*	*5.2*
γ_4	-0.06 ***	-0.12 ***	-0.14 ***	-0.08 ***	-0.11 ***
	-3.3	*-5.1*	*-7.6*	*-4.1*	*-4.9*

Consumer Services

	FD&DG RET	GEN RET	MEDIA	TRA&LEI	
γ_1	-0.22 ***	-0.17 ***	-0.14 ***	-0.17 ***	
	-4.1	*-3.8*	*-3.8*	*-3.2*	
γ_2	0.97 ***	0.98 ***	0.98 ***	0.98 ***	
	161.7	*190.8*	*221.2*	*155.7*	*blank*
γ_3	0.13 ***	0.14 ***	0.17 ***	0.17 ***	*space*
	5.4	*5.0*	*6.9*	*6.0*	
γ_4	-0.08 ***	-0.11 ***	-0.09 ***	-0.08 ***	
	-4.4	*-5.2*	*-5.0*	*-3.6*	

Table excludes certain subindices of industrials: IND TRA, SUP SER and consumer goods: PER GDS, TOBACCO due to limited space. EGARCH estimation for excluded subindices is fully comparable to displayed indices and data is available upon request. *** denotes statistical significance at the 1% level. t-statistics in italic.

Table A6.2: EGARCH (1,1) Estimation of selected Financial Stocks for 2007-2010

	REITs (N=119)			Asset Managers (N=30)		
By γ_4	Min	Medain	Max	Min	Medain	Max
	ANWORTH MGE.ASSET	EQUITY ONE	ANTHRACITE CAP.	AFFILIATED MANAGERS	FRANKLIN RESOURCES	INNOVARO
γ_1	-0.34 ***	-0.08 ***	-0.56 ***	-0.12 ***	-0.09 ***	-0.26 ***
	-7.7	*-2.8*	*-7.7*	*-4.0*	*-2.9*	*-9.3*
γ_2	0.95 ***	0.99 ***	0.79 ***	0.98 ***	0.99 ***	0.96 ***
	147.7	*201.4*	*45.4*	*200.8*	*189.4*	*203.6*
γ_3	0.26 ***	0.12 ***	1.13 ***	0.18 ***	0.13 ***	0.28 ***
	12.1	*6.2*	*37.8*	*6.8*	*6.4*	*15.5*
γ_4	-0.23 ***	-0.07 ***	0.54 ***	-0.15 ***	-0.07 ***	0.06 **
	-15.0	*-3.6*	*27.7*	*-6.9*	*-3.2*	*2.4*

	Banks (N=119)			Brokers (N=19)		
By γ_4	Min	Medain	Max	Min	Medain	Max
	CEN. JER-SEY BANC.	ESSA BANCORP	UTD. BANCORP OH.	MARKETAXESS HOLDINGS	SWS GP.	PENSON WORLDWIDE
γ_1	-0.15 ***	-0.26 ***	-0.11 ***	-0.14 ***	-0.02 **	-0.18 ***
	-8.9	*-6.2*	*-5.3*	*-3.8*	*-2.3*	*-3.9*
γ_2	0.98 ***	0.97 ***	0.98 ***	0.98 ***	1.00 ***	0.97 ***
	143.1	*128.2*	*194.0*	*156.5*	*402.1*	*105.6*
γ_3	-0.01	0.17 ***	0.16 ***	0.20 ***	0.12 ***	0.30 ***
	0.2	*7.8*	*11.3*	*8.4*	*6.3*	*9.7*
γ_4	-0.38 ***	-0.07 ***	0.02 ***	-0.12 ***	-0.06 ***	0.05 ***
	-39.2	*-2.7*	*1.9*	*-6.3*	*-3.9*	*3.0*

	Insurers (N=61)					
By γ_4	Min	Medain	Max			
	AFLAC	PRUDENTIAL FINL.	MERCER INSUR. GP.			
γ_1	-0.13 ***	-0.04 **	-0.35 ***			
	-3.9	*-2.0*	*-6.2*			
γ_2	0.98 ***	0.99 ***	0.95 ***		*blank*	
	203.9	*316.8*	*128.4*		*space*	
γ_3	0.33 ***	0.21 ***	0.28 ***			
	15.2	*9.1*	*15.4*			
γ_4	-0.16 ***	-0.08 ***	0.05 ***			
	-6.7	*-3.9*	*2.9*			

Stocks displayed are exemplarily selected for each subsample of financial stocks. *$\gamma 4$* as coefficient regarding leverage effect serves as reference for stock selection. Full data is available upon request. ***, **, and * denote statistical significance at the 1%, 5%, and 10% level respectively.

Table A6.3: Descriptive Statistics on Financial Industry Subindices

07/2008 - 07/2010[1]	*FTSE NA Financials*	FTSE NA REIT	FTSE NA BANKS	NA-DS Insurance	NA-DS Asset Mgrs.	US-DS Insur. Brokers
Daily Returns						
Mean	*-0.05%*	-0.03%	-0.02%	-0.05%	-0.05%	-0.03%
Median	*0.00%*	0.00%	0.00%	0.08%	0.08%	0.01%
Skewness	*-0.12*	-0.09	0.06	-0.18	-0.27	-0.06
Kurtosis	*6.19*	5.66	6.77	5.78	7.58	6.39
StD	*0.04*	0.04	0.04	0.03	0.04	0.02
Q_{10}	*216*[a]	354[a]	166[a]	304[a]	145[a]	187[a]
Q_{100}	*901*[a]	1164[a]	764[a]	986[a]	659[a]	597[a]
$ARCH_{10}$	*44*[a]	83[a]	27[a]	49[a]	8[b]	85[a]
$ARCH_{100}$	*168*[a]	151[a]	172[a]	170[a]	154[a]	159[a]
Daily Variance						
Mean	*.0012*	.0019	.0017	.0007	.0013	.0004
Median	*.0006*	.0008	.0008	.0004	.0007	.0003
Skewness	*1.40*	1.42	1.13	1.89	1.50	1.33
Kurtosis	*4.48*	4.41	3.29	6.60	4.51	4.05
StD	*.0012*	.0020	.0017	.0007	.0014	.0004
Q_{10}	*19*[a]	10[b]	14[b]	32[a]	12[b]	5[b]
Q_{100}	*85*[b]	91[b]	77[b]	124[b]	103[b]	87[b]
$ARCH_{10}$	*27*[a]	10[b]	14[b]	38[a]	1[b]	1[b]
$ARCH_{100}$	*80*[b]	128[a]	61[b]	164[a]	63[b]	34[b]
Correlation						
Mean	*0.90*	0.84	0.83	0.89	0.90	0.72
Median	*0.90*	0.84	0.82	0.89	0.91	0.74
Skewness	*-0.49*	-0.16	0.28	-0.04	-0.90	-0.90
Kurtosis	*3.16*	3.27	2.18	2.64	3.11	2.69
StD	*0.01*	0.02	0.02	0.02	0.01	0.07
Beta						
Mean	*1.55*	1.82	1.70	1.14	1.64	0.77
Median	*1.52*	1.74	1.61	1.14	1.61	0.73
Skewness	*0.52*	0.72	0.72	0.14	0.60	0.93
Kurtosis	*2.76*	3.12	2.90	2.78	3.18	5.57
StD	*0.32*	0.51	0.44	0.22	0.30	0.22

[1]) Descriptive mean stock return statistics on the basis of 504 trading days throughout the period of the financial crisis under investigation; Squared stock returns are tested by Ljung-Box test for serial correlation ($Q_{10/100}$) and by ARCH-test for heteroscedasticity for 10 and 100 lags; [a]) indicates test result is positive for serial correlation or heteroscedasticity at the 5% significance level, [b]) indicates the opposite.

Table A6.4: Correlation Adjustments of FSI and Other Industries around Bank Failures and Bailouts

Event wind.	FSI (nobs=35) Mean	FSI (nobs=35) Median	IGI (nobs=49) Mean	IGI (nobs=49) Median	CGI (nobs=49) Mean	CGI (nobs=49) Median	CSI (nobs=28) Mean	CSI (nobs=28) Median
Bank Failures								
0;0	-.0001	-.0001	-.0009 ***	-.0004 **	-.0018 **	-.0006 **	-.0015 *	-.0007 **
1;1	.0011	.0000	-.0004	-.0004	-.0015	-.0005 *	-.0009	-.0022 **
2;2	.0007	.0001	-.0003	-.0010	.0006	-.0008	-.0007	-.0026
1;-1	-.0002	-.0004	-.0002	-.0001	-.0005	.0000	-.0010 *	-.0004
1;0	-.0003	-.0008	-.0011 **	-.0004 **	-.0023 **	.0002 **	-.0025 ***	-.0008 ***
2;0	.0000	-.0001	-.0008	-.0005 *	-.0003	-.0004	-.0016	-.0010
0;1	.0013	.0006	-.0002	-.0003	-.0010	-.0009	.0001	-.0010
0;2	.0007	-.0001	-.0003	-.0004	-.0010	-.0012	-.0007	-.0011 **
Bank Bailouts								
0;0	.0003	.0003 **	.0003 ***	.0002 ***	.0004 *	.0004 *	.0003	.0003 ***
1;1	.0020 *	.0001	.0016 ***	.0004 **	.0031 ***	.0005 ***	.0023 ***	.0006 ***
2;2	.0019 **	.0010 ***	.0013 ***	.0004 ***	.0023 **	.0010 **	.0032 ***	.0022 ***
1;-1	.0004	.0000	.0003	.0000	.0005	-.0001	.0003	.0001
1;0	.0007 *	.0002 *	.0006 **	.0002 *	.0009 *	.0008 *	.0006	.0003 *
2;0	.0009	.0013 **	.0008 **	.0004 ***	.0005	.0009	.0017 ***	.0012 ***
0;1	.0016 *	.0001	.0013 ***	.0005 ***	.0026 ***	.0005 ***	.0020 ***	.0005 ***
0;2	.0013	.0002	.0009 **	.0004 **	.0022 **	.0003 **	.0018 ***	.0006 ***

Nobs: number of observations for each sample, failures and bailouts - calculated as number of stocks multiplied each by seven failure and seven bailout dates; ***, **, and * indicate statistical significance at the 1%, 5%, and 10% levels, respectively; applied test statistics are the Wilcoxon signed rank test as non-parametric median test and the standard t-test as parametric mean test. FSI: Financial Services industry; IGI: Industrial Goods industry; CGI: Consumer Goods industry; CSI: Consumer Services industry.

Table A6.5: Beta Adjustments of FSI and Other Industries around Bank Failures and Bailouts

Event	FSI (nobs=35)		IGI (nobs=49)		CGI (nobs=49)		CSI (nobs=28)	
wind.	Mean	Median	Mean	Median	Mean	Median	Mean	Median
Bank Failures								
0;0	0.011	0.007	0.012 *	0.014 ***	0.000	0.003	0.013	0.011 *
1;1	-0.016	-0.005	-0.009	0.015	-0.029 ***	-0.030 ***	-0.015	-0.011
2;2	0.022	0.045	-0.021 *	-0.014	-0.037 ***	-0.053 ***	-0.010	0.007
1;-1	-0.003	0.005	0.004	0.011 ***	-0.004	0.003	0.002	0.008
1;0	0.007	0.018	0.016 *	0.023 **	-0.004	-0.005	0.015	0.016
2;0	0.018	0.031	0.002	-0.001	-0.011	-0.005	0.019	0.013
0;1	-0.012	-0.009	-0.014	0.010	-0.025 ***	-0.016 ***	-0.018	0.000
0;2	0.014	-0.001	-0.011	0.010	-0.025 ***	-0.022 ***	-0.016	-0.010
Bank Bailouts								
0;0	0.001	0.006	0.004	0.007	0.002	0.006	0.007	0.011
1;1	-0.005	0.025	0.011	0.035 *	0.004	0.010	0.026 *	0.032 *
2;2	0.019	0.005	0.024	0.038 **	0.014	0.019	0.041 **	0.027 **
1;-1	-0.019	-0.013	0.002	0.009	-0.007	0.010	0.003	0.007
1;0	-0.019	-0.024	0.006	0.016	-0.006	-0.001	0.010	0.020
2;0	0.008	-0.018	0.025	0.033 **	0.011	0.009	0.026 *	0.010
0;1	0.015	0.013	0.009	0.017	0.011	0.010	0.023 *	0.032 *
0;2	0.012	0.016	0.003	0.015	0.005	0.004	0.022 *	0.034

Nobs: number of observations for each sample, failures and bailouts - calculated as number of stocks multiplied each by seven failure and seven bailout dates; ***, **, and * indicate statistical significance at the 1%, 5%, and 10% levels, respectively; applied test statistics are the Wilcoxon signed rank test as non-parametric median test and the standard t-test as parametric mean test. FSI: Financial Services industry; IGI: Industrial Goods industry; CGI: Consumer Goods industry; CSI: Consumer Services industry.

Table A6.6: Multivariate Regressions on FSI's and the S&P 500's Variance Shifts around Bank Events

		Financial Services Industry						S&P 500 Composite			
Coefficients		Model 1		Model 2		Model 3		Model 4			
Intercept	*δ1*	(0.01)		0.01		0.00		0.02		0.02	
USD failure assets	*δ2*			(0.00)		(0.00)				(0.00)	***
USD bailout capital	*δ3*			(0.00)		(0.00)				0.00	
t-1	*δ4*			(0.00)							
t	*δ5*	0.01						(0.01)			
t+1	*δ6*					0.01				(0.00)	
Wednesday	*δ7*	0.01		(0.05)	***	(0.05)	***	(0.08)		(0.09)	
S&P500	*δ8*	0.59	***	0.58	***	0.58	***				
FSI_{t-1}, $S\&P_{t-1}$	*δ9*	(0.07)	**	(0.07)	**	(0.07)	**	(0.13)	**	(0.13)	**
Nobs.		205		205		205		41		41	
F		0.00		0.00		0.00		0.72		0.84	
adj. R^2		0.11		0.11		0.11		-0.01		-0.05	
DW		1.99		2.05		2.05		2.14		2.18	

(x): values in parentheses are negative, Nobs: number of observations - 3 trading days (*t-1, t, t+1*) per event (7 failures and seven bailouts) and index (5 subindices for financial services). ***, **, and * indicate statistical significance at the 1%, 5%, and 10% levels, respectively.

Corporate Finance and Governance

Herausgegeben von Dirk Schiereck

Band 1 Sebastian Michael Gläsner: Return Patterns of German Open-End Real Estate Funds. An Empirical Explanation of Smooth Fund Returns. 2010.

Band 2 Patrick Ams: Directors' Dealings and Insider Trading in Germany. An Empirical Analysis. 2010.

Band 3 Joachim Vogt: Value Creation within the Construction Industry. A Study of Strategic Takeovers. 2011.

Band 4 Fabian Braemisch: Underpricing, Long-Run Performance, and Valuation of Initial Public Offerings. 2011.

Band 5 Matthäus Markus Sielecki: Creating and Governing an Integrated Market for Retail Banking Services in Europe. A Conceptual-Empirical Study of the Role of Regulation in Promoting a Single Euro Payments Area. 2011.

Band 6 Arne Wilkes: Determinants of Credit Spreads. An Empirical Analysis for the European Corporate Bond Market. 2011.

Band 7 Dirk Schiereck / Martin Setzer (Hrsg.): Bankerfolg und Akquisitionen. Aktuelle Erkenntnisse zur Konsolidierung und Restrukturierung der Finanzindustrie. 2011.

Band 8 Steffen Meinshausen: M&A Activity, Divestitures and Initial Public Offerings in the Fashion Industry. 2012.

Band 9 Malte Helmut Raudszus: Financial Return Risk and the Effect on Shareholder Wealth. 2012.

www.peterlang.de

www.ingramcontent.com/pod-product-compliance
Ingram Content Group UK Ltd.
Pitfield, Milton Keynes, MK11 3LW, UK
UKHW020835290926
15470UKWH00004B/1369

9 783631 622490